Important Bird Areas of New York

Important Bird Areas of New York
by Michael F. Burger and Jillian M. Liner

200 Trillium Lane
Albany, NY 12203

http://ny.audubon.org

Book design, cover design, and layout by Christi A. Sobel
Copy edited by Elissa Wolfson
Printed at BookMasters, Inc.

Front cover photos © Marie Read (from top to bottom):
Scarlet Tanager
Least Bittern
Piping Plover
Bobolink
Back cover photo © Carl Heilman II / Wild Visions, Inc.
Panoramic view of Adirondack High Peaks

ISBN 0930698371

Important Bird Areas *of* New York

Second Edition

Habitats Worth Protecting

Michael F. Burger and Jillian M. Liner

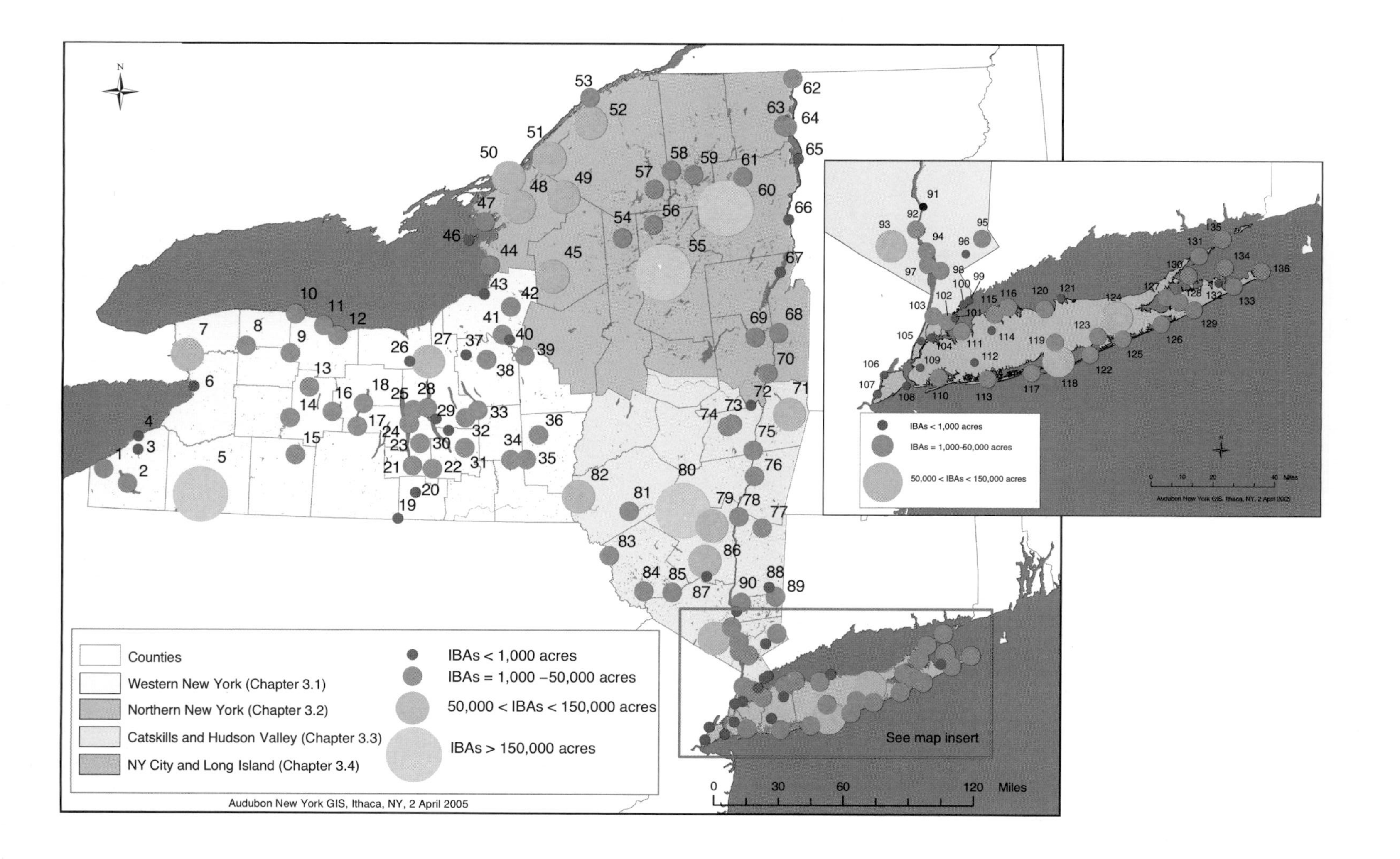

N
Counties
Western New York (Chapter 3.1)
Northern New York (Chapter 3.2)
Catskills and Hudson Valley (Chapter 3.3)
NY City and Long Island (Chapter 3.4)
IBAs < 1,000 acres
IBAs = 1,000 –50,000 acres
50,000 < IBAs < 150,000 acres
IBAs > 150,000 acres
See map insert
0 30 60 120 Miles
Audubon New York GIS, Ithaca, NY, 2 April 2005
IBAs < 1,000 acres
IBAs = 1,000-50,000 acres
50,000 < IBAs < 150,000 acres
0 10 20 40 Miles
Audubon New York GIS, Ithaca, NY, 2 April 2005

Dear Reader,

It is my pleasure to introduce the 2nd edition of *Important Bird Areas of New York*. This edition builds upon Audubon New York's well-earned reputation as one of the premier conservation organizations in our state. Audubon New York has been at the forefront of conservation excellence during our administration and has been a partner in numerous environmental success stories including protecting our state's birds from threats posed by lead sinkers, harmful pesticides, and invasive species.

In addition, Audubon played an integral role in working with the state on the Bird Conservation Areas (BCA) Law, which New York modeled after Audubon's Important Bird Areas program. The BCA program in New York was the first of its kind in the nation and established a process to identify important bird habitat on state-owned lands and waters. To date, New York has designated twenty-eight Bird Conservation Areas and plans to add twenty more by 2007. It is also notable that Audubon New York's IBA program, launched in 1996, and its first IBA book, published in 1998, now serve as models for 45 other states that now have IBA programs.

This edition also is particularly useful to active birders as it identifies which IBAs provide the best birding opportunities. As an avid birder, I can personally attest to the warm and enduring memories associated with birding with my family and friends in the great outdoors.

Celebrating its centennial anniversary in 2005, Audubon has made a real difference in the quality of life throughout our state and the entire nation by means of education, science and advocacy. Happy birding, and enjoy the book.

Sincerely,

George Pataki
Governor, New York State
April 2005

Table of Contents

Acknowledgements

This book would not have been possible without assistance from numerous individuals, whom the authors sincerely thank. We've done our best to thank everyone who helped with this endeavor, but if we missed anyone, please accept our sincere apologies.

At the very heart of the Important Bird Areas identification process was the IBA Technical Committee, which consisted of the following members: Elliott Adams, Oakes Ames, Tim Baird, Bob Budliger, Valerie Freer, John Fritz, Peter Gibbs, Jane Graves, Jay Greenberg, Lee Harper, Sheila Hess, Eric Lind, Kevin McGowan, Marty McGuire, Ralph Odell, Bill Ostrander, Ray Perry, Mike Richmond, Ken Rosenberg, Seymour Schiff, Bryan Swift, and David VanLuven. The contribution that this group made would be difficult to overstate. They gave freely of their time to attend meetings and conference calls, oversee the process, review site nominations, provide data, and decide which sites warranted IBA status. Throughout the process they repeatedly demonstrated their ability to weigh the many challenges and options we faced, and, in the end, provide solid, practical advice for how to move forward. Their unwavering commitment to the IBA concept and the integrity of the IBA designation have provided Audubon New York with a solid network of sites upon which to focus its conservation efforts. Thank you.

We also wish to thank the staff of Audubon New York, especially Richard Anderson, Albert Caccese, Graham Cox, Richard Haley, Jamie Halperin, Julie Hart, Mitschka Hartley, Eric Lind, Sean Mahar, Richard Merritt, David Miller, Michael Morgan, Thomas O'Handley, Glenn Phillips, and Aimee Tweedie. They supported this effort in so many ways, including reviewing drafts of this text, inputting data into the database, conducting field surveys, coordinating outreach, and simply believing that this project was important to our organization. Working with this group of dedicated and capable individuals is truly an honor.

Our colleagues in the implementation of IBA programs contributed to this effort through their leadership, wise counsel, and support. In particular, we thank Jeff Wells, who coordinated the first round of identifying IBAs, compiled the first edition of *Important Bird Areas in New York State*, and laid a strong foundation for New York's IBA program. Similarly, the staff of Audubon's national science office, especially Greg Butcher, Frank Gill, Tess Present, and Connie Sanchez, strengthened that foundation by providing a virtual clearinghouse of good ideas and by developing a close partnership with BirdLife International, as did John Cecil and Dan Niven, who also reviewed sections of this manuscript. Lincoln Fishpool of BirdLife International helped by emphasizing the importance of adhering to standard criteria and sharing approaches from around the world for both identifying and de-listing IBAs. The strong network of Audubon chapters in New York also assisted in numerous aspects of the IBA identification process, including nominating sites, reviewing site summaries, providing bird data, and assisting in the conservation of IBAs; many individual chapter members are listed below. Additionally, we thank our fellow state IBA coordinators for discussing issues and potential solutions, confirming best approaches based on their experiences, and commiserating on the enormity of certain tasks.

Numerous others contributed in countless ways, including nominating sites, supplying data, and reviewing site summaries. For this, we thank Lorie Alcaide, Larry Alden, Sharon Anderson, Robert Andrle, John Askildsen, Lynn Barber, Howard Barton, Douglas Bassett, Mona Bearor, Alan Benton, Steven Biasetti, Michael Bochnik, Jeff Bolsinger, Howard Bolston, David Bonter, Wendy Borden, Ronald Bourque, Joseph Brin, Greg Brown, Bill Brown, Dirk Bryant, Carol Capobianco, Bernie Carr, Fred Caslick, John Cecil, Lance Clark, Geoffrey Ryan, Jeremy Coleman, Joan Collins, John Confer, Paul Connor, Nick Conrad, William E. Cook, Patrick Crast, Scott Crocoll, Peter Davidson, Leonard De Francisco, Anthony Deobil, Mickey Dietrich, Bob Donnelly, Barbara Drake, James Eckler, Karen Edelstein, Nancy Engle, Mike Ermer, James Farquhar, Larry Federman, Michael Feller, Ken Feustel, Emily FitzGerald, Dave Forness, Marcia Fowle, Jack Frocht, Lucy Gentile, Michael Gibbons, Linda Gibbs, Neil Gifford, Tracy Gingrich, Andrew Goeller, William Gorman, Eric Grace, John Gregoire, Mark Gretch, David Grove, Sharon Gunkel, Sunita Halasz, Stefan Hames, John Hannon, Helen Hays, Ed Henry, John Hersey, Paul Hess, Kevin Holcomb, John Jablonski III, Florence James, Tom Jasikoff, Glenn Johnson, Steve Kahl, Mark Kandel, Paul Keim, Jim Kimball, Elizabeth King, Erik Kiviat, Jason Klein, Ken Kogut, Mary Ellen Kris, Bill Krueger, David Künstler, Tom LeBlanc, Geoff LeBaron, Gary Lee, Daniel Levy, Matt Levy, Eric Liner, Robert Long, Ruth Lundine, Ted Mack, Francine Maftrantuono, Bruce Manuel, Robert Marcotte, Irene Mazzocchi, E.J. McAdams, Kevin McGann, Hugh McGuinness, Judy McIntyre, Lori McKean, Bill McKeever, Dick Miga, Joanne Mitchell, Jessica Morgan, Mike Morgante, Terry Mosher, John Moyle, Mary Mulcahy, Frank Murphy, Richard Nelson, Dan Niven, Dave Odell, Peter O'Shea, John Ozard, Drew Panko, Bob Parris, Karl Parker, Bonnie Parton, Sean Prockter, Bruce Penrod, John M. Peterson, Marie Petuh, William Purcell, Barb Putnam, Glenn Quinn, Joe Racette, Ray Rainbolt, Gill Randell, Zoe Richards, Don Riepe, Gerald Rising, Jack Robbins, Donald Root, Richard Rosche, Ken Ross, Tom Salo, Cathie Sandell, Mike Scheibel, Mickey Scilingo, Kelly Sheridan, Dominic Sherony, Gerry Smith, Bob Spahn, Michael Sperling, Scott Stoner, David Suggs, June Summers, Scott Sutcliffe, Chris Tessaglia-Hymes, John Thompson, Molly Thompson, Doug Thompson, James Utter, Patricia Vissering, Karen Wallace, Mike Wasilco, Ann Watson, Norm Webber, Carol Weiss, Stephanie Weiss, Alan Wells, Della Wells, Jeff Wells, Linda White, Jennifer Wilson-Pines, Anne Wong, Amie Worley, Matthew Young, and Andrew Zepp.

Several individuals, programs, and organizations provided additional data from their databases, allowing this effort to be as comprehensive as possible, including the Adirondack Cooperative Loon Program (Nina Schoch), Bird Studies Canada's Marsh Bird Monitoring (Kathy Jones); Audubon and the Cornell Lab of Ornithology's eBird; Manomet's International Shorebird Survey (Brian Harrington); New York City Audubon's Harbor Herons Project (E.J. McAdams and Yigal Gelb); New York State Department of

Environmental Conservation's Endangered Species Unit (Pete Nye and Barbara Loucks), New York Natural Heritage Program (David VanLuven and Nick Conrad); New York State Breeding Bird Atlas (Kim Corwin); New York State Department of Environmental Conservation's Nongame and Habitat Unit (Bryan Swift, Tim Post, and Dave Adams); New York State Office of Parks, Recreation & Historic Preservation (Tom Lyons, Ray Perry, and Christina Ricci); New York State Ornithological Association (Tim Baird, Kevin McGowan, and Bryan Swift); and Vermont Institute of Natural Science (Dan Lambert). And for creating a place to organize and store all of those data, thanks to volunteer Jim Roble who redesigned our database, greatly facilitating the site identification process.

Our sincere thanks to John Fitzpatrick for hosting Audubon New York's bird conservation program at the incomparable Cornell Laboratory of Ornithology. The Lab of O's supportive and collaborative environment, in which ideas and advice are freely expressed, and its host of resident experts with whom we regularly interact, helped improve this effort in many ways. We especially thank Tom Fredericks, Kevin McGowan, Ken Rosenberg, Roger Slothower, Diane Tessaglia-Hymes, and Marge Villanova.

Finally, we thank the professionals who assisted with the completion of this book, including our designer, Christi Sobel; copy editor, Elissa Wolfson; those who supplied us with images, including Patty Coan, the Cornell Lab of Ornithology, Bill Dyer, Michael Gambino, Carl Heilman II, Eric Lind, Jerry and Sherry Liguori, Jim Logan, Jeff Nadler, Marie Read, Brandt Ryder, Brian Sullivan, Mary Tremaine, and Gerrit Vyn; and Audubon employees Dan Cooper, Jim Logan, and Nancy Severance, who helped with various aspects of the publication process.

We look forward to continuing our work with all of you to protect these critical sites.

Mike Burger
Jillian Liner

The authors wish to extend their gratitude to the leadership of Audubon for having the vision to recognize the significance of the Important Bird Areas program to the pursuit of Audubon's mission of protecting birds, other wildlife, and their habitats; for their commitment to science-based programs; and for their support of IBAs at the international, national, state, and local levels. In particular, we thank:

National Audubon Society's Board of Directors
Chairs: Carol Browner, Donal O'Brien
President: John Flicker

Audubon New York's Board of Directors
Chair: Constantine Sidamon-Eristoff
Executive Director: David J. Miller

Audubon Council of New York State
Chairs: Gladys Goldmann, Geoffrey Cobb Ryan

Special thanks also go to the following institutions that directly supported the second round of IBA identification, and the publication of this book:

Sarah K. deCoizart Charitable Trust
Ellsworth Kelly Foundation, Inc.
New York State Biodiversity Research Institute

Audubon New York wishes to acknowledge the following institutions and individuals for their generous financial support of Audubon New York and the New York IBA program, without which completion of this book would not have been possible:

Harriet Ford Dickenson Foundation
Margot P. Ernst
Graham Challenge Fund
Marian S. Heiskell
The Henry Luce Foundation, Inc.
LuEsther T. Mertz Charitable Trust
The New York Community Trust
Park Foundation, Inc.
Howard Phipps Foundation
Constantine Sidamon-Eristoff
Virginia K. Stowe

Foreword

The National Audubon Society can trace its birth to actions taken right here in New York in January 1905. At that time, many people saw nature as sustainable and without limits. However, as fashion and the desire for feathers threatened bird populations, a group of concerned citizens in the Northeast recognized the need to protect these magnificent creatures. Working through emerging Audubon chapters across the country, the nascent organization successfully lobbied for a series of bird protection laws on the state and federal levels. From this model, the National Audubon Society and the modern conservation movement grew.

Today, the National Audubon Society includes more than 500 Audubon chapters, a growing national network of Audubon Centers and close to a half million members. We have succeeded by adhering to our mission: to protect birds, wildlife, and their habitat for the benefit of all life, and a principle: that people acting together can meet any challenge.

The National Audubon Society thrives through the grassroots action of individuals in their communities. We call it Citizen Science, and examples include our Christmas Bird Count and Great Backyard Bird Count. These programs bring Chapter members and committed individuals out into the field to count birds and learn about conservation firsthand.

The cornerstone of Audubon's conservation efforts centers on our Important Bird Areas (IBA) program. Here, Audubon's professional and "citizen" scientists fan out across the nation to identify Important Bird Areas (IBAs) – land, that based on scientific data – is deemed crucial to birds' survival. To date, Audubon has identified over 1,600 sites in 46 states, encompassing both public and private lands. IBAs are connecting all the places that matter to maintaining healthy bird populations, and to maintaining the conservation of all biodiversity.

New York started its IBA program in 1996, and published its first IBA book in 1998. This second edition couldn't have come at a better time. Audubon's 2004 "State of the Birds" report finds nearly 30% of more than 400 species tracked by the Breeding Bird Survey have declined significantly. The loss of native habitat and habitat fragmentation are seen as leading causes of this decline. Since birds serve as sentinels of our environment's health, we need to continue monitoring their health carefully.

This second edition IBA book provides important information to assist in bird habitat restoration efforts and will serve as an important resource for decision makers at every level of government, as well as developers, planners, educators, property owners and conservation organizations.

My sincere congratulations to David J. Miller, Executive Director of Audubon New York, and his remarkable bird conservation team ably led by Michael Burger and Jillian

Liner, the authors of this important book, for pulling together a coalition of ornithological and birding leaders and top level conservation biologists to identify these Important Bird Areas. This book will serve as a model for other states as Audubon continues its mission to protect birds, other wildlife, and their habitats across the country.

John Flicker, President
National Audubon Society
April 2005

Chapter I
Introduction

© Carl Heilman II / Wild Visions, Inc.

Delaware River

Habitat is the key to conservation. To thrive, all species need the right kinds of places to carry out the business of living and reproducing. Without those places, populations decline and, in the most severe cases, are extirpated or go extinct. Even though they are more mobile than many other organisms, birds are not exempt from this habitat requirement. Without adequate and appropriate places to nest, forage, rest during migration, and over-winter, bird populations decline. It is well known to conservation biologists that habitat loss and degradation are the leading threats to vulnerable bird species in North America. It follows that habitat protection and proper stewardship are the primary goals of bird conservationists.

But not all habitats are of equal value for conservation. Clearly some habitats are more important than others for sustaining populations of native birds. For example, habitats that are degraded or heavily impacted by humans rarely support significant numbers of birds of conservation concern. Such habitats often harbor non-native species that out-compete native birds for nest sites and other resources. But places that support species of conservation concern or large populations of native birds are vital to the conservation of birds. Those sites—important bird areas—are the subject of this book.

The Important Bird Areas (IBA) program is a bird conservation initiative with simple goals: to identify the most important places for birds, and to conserve them. IBAs are identified according to standardized, scientific criteria through a collaborative effort among state, national, and international non-governmental conservation organizations (NGOs), state and federal government agencies, local conservation groups, academics, grassroots environmentalists, birders, and others. As a result, IBAs link global and continental bird conservation priorities to local sites that provide critical habitat for native bird populations.

After sites have been identified, the IBA program serves as a catalyst for achieving their conservation, often with the same partners. The tools employed to achieve conservation of IBAs include conservation planning, science-based habitat stewardship and management, and site protection. These activities encompass a broad array of scientific, educational, and advocacy initiatives on local, state, and national levels, including open space protection, smart growth programs, habitat restoration, bird monitoring and censuses, and public and private landowner education. IBAs truly have become a global conservation currency—they are a widespread and accepted method to focus bird conservation work.

In this book, you will find a brief history of the Important Bird Areas program, information about how sites are identified as IBAs, and how IBAs fit into the big picture of bird conservation. However, the most significant contents of this book are the descriptions of the IBAs of New York State, including detailed information about why they were identified and suggestions for ways you can help to promote their conservation. The future of New York's birds will depend in large part upon the protection and proper management of these sites.

History of Important Bird Areas

BirdLife International (hereafter called BirdLife), a global partnership of environmental organizations and research institutions, identified the first IBAs in Europe in the mid-1980s. Using standardized criteria, sites of global importance for the conservation of birds were selected. The idea of using these criteria for site identification subsequently spread across the globe, with programs initiated by BirdLife partners in the Middle East, Africa, and Asia. In the mid-1990s, the IBA concept arrived in the Americas. At that time, the National Audubon Society and the newly created American Bird Conservancy (ABC, then partner to BirdLife) together adopted the IBA program in the U.S., with Audubon establishing programs at the state level and ABC addressing the national picture. In 1995, Audubon launched the Pennsylvania IBA program, followed in 1996 with the start of the New York IBA program. Also in 1996, the Canadian and Mexican IBA programs were begun. In New York, more than 250 sites were nominated during that first round of identifying IBAs. After review by a technical committee composed of 22 bird experts from around the state, 127 sites were identified as IBAs and announced with the publication of *Important Bird Areas in New York State* (Wells 1998). New York's was the first IBA program in the Americas to complete the site identification phase. Currently, IBA programs exist in 130 countries around the world, including 21 countries in the Americas. In the U.S., 46 states have IBA programs.

Conservation successes began almost immediately for Audubon New York's IBA program, and they continue today. Over the ensuing years, numerous IBAs were protected through acquisition or purchase of conservation easements by state and federal governments, land trusts, and other NGOs. Lobbying efforts by Audubon and its conservation partners helped increase funding for state and federal land conservation programs and directed funds toward these priority sites. Effective public education and grassroots support from local Audubon chapters and others also played important roles in these successes. Additional successes and activities included habitat restoration projects, conservation planning at high-profile sites, coordinated bird monitoring at several IBAs, and numerous public outreach endeavors. One especially notable success was the establishment of the New York State Bird Conservation Area (BCA) program through legislation passed in 1997—the first state BCA law in the country. This program identifies state-owned properties critical for bird conservation and sets up a process to prepare and implement management plans (see box 1.1). Several years later, Westchester County enacted the first county-wide BCA law in the country.

Early in the new 21st century, Audubon began working more closely with BirdLife, and in 2004 the organization became the official U.S. partner for the implementation of IBAs. During this period, Audubon hired national science staff to assist and coordinate state IBA programs, and to work directly with BirdLife to identify global and continental level IBAs in the U.S. To oversee this latter task, a National IBA Technical Committee was convened (see Chapter 2). For several years, Audubon has organized an IBA conference for the Americas, where IBA leaders gather to exchange ideas and plan collaborative efforts.

At the same time that Audubon was escalating its IBA efforts, significant progress was being made across North America on major continental-scale bird conservation initiatives, including those for landbirds (Rich *et al.* 2003), waterfowl (NAWMP Committee 1999), waterbirds (Kushlan *et al.* 2002), and shorebirds (Brown *et al.* 2000). Under the umbrella of the North American Bird Conservation Initiative (NABCI), collaborative workshops were organized around Bird Conservation Regions (BCRs) devised by NABCI for that purpose (U.S. NABCI Committee 2000). In many cases, IBAs were identified as focus areas for implementing collaborative habitat conservation projects, and they were formally adopted as one of the habitat conservation strategies that make up the Partners in Flight (PIF) bird conservation strategy.

When the IBA program was initiated in New York in 1996, Audubon determined that the site identification process should be repeated every five to ten years to make sure that the network of sites remains as complete as possible and that the most recent bird data are taken into consideration. By 2002, three additional factors were pointing toward the need to launch a second round of identifying IBAs in New York. First, Audubon's increased ties with BirdLife made it apparent that the state criteria needed to be more fully aligned with the global criteria. Second, the opportunity existed for IBAs to better serve as focus areas for the bird conservation initiatives under the NABCI umbrella by integrating priority species (especially those of PIF) into the IBA criteria. And, third, the ability to perform a comprehensive, statewide approach to IBA identification was made possible by the acquisition of relatively recent Geographic Information Systems (GIS) technology and data layers. For all of these reasons, Audubon New York embarked upon its second round of identifying IBAs in 2002, reaching out to partner organizations such as local Audubon chapters, other conservation organizations and agencies, and academic institutions to participate in the process.

Details about the second round process and criteria are in Chapter 2, and summaries of the sites identified as IBAs can be found in Chapter 3.

Box 1.1

New York State Bird Conservation Areas

Based on legislation signed by Governor George E. Pataki in 1997, the State of New York established a Bird Conservation Area (BCA) program, the first of its kind in the nation. Modeled after Audubon's IBA program, the BCA program seeks to identify the most important state-owned bird habitats and to integrate bird conservation needs into the management of those sites, within the context of agency missions. Critical bird breeding, migratory stopover, feeding, and over-wintering areas are identified according to criteria patterned after Audubon's IBA criteria and set forth in law. These criteria pertain to sites with significant native bird concentrations; sites with endangered, threatened, or special concern species; sites that are examples of representative or unique habitats; and sites important for bird research and long-term monitoring.

Currently, the BCA program employs four full-time agency biologists who work with local site managers to identify sites, complete management guidance summaries, and plan research, management, and interpretive projects for the sites. Oversight of the BCA program, as set forth in law, is provided through a BCA Program Advisory Committee, which consists of representatives of state agencies, and environmental and ornithological groups.

As of this writing, 28 BCAs have been designated, comprising more than 190,000 acres of habitat (see list below). These sites cover a diversity of habitats across the state, including wetlands along the Hudson and Niagara Rivers, Lake Ontario, and Long Island Sound; forests in Central New York and along the Hudson River where Cerulean Warblers nest; and high elevation forests in the Adirondack and Catskill Mountains that are home to the Bicknell's Thrush. In January 2005, Governor Pataki, during his State of the State message, pledged the creation of 20 new BCAs to add to the 28 already designated. Designation of BCAs helps integrate the proper management of these critical bird habitats into the planning processes for these public lands.

See the New York State Bird Conservation Area website for more information (http://www.dec.state.ny.us/website/dfwmr/wildlife/bca/).

New York State Bird Conservation Areas
name; size; location; and date of designation

1. David A. Sarnoff Pine Barrens Preserve; 2,324 acres; Suffolk County; 31 August 1998.
2. Eastern Lake Ontario Marshes; 4,940 acres; Oswego and Jefferson Counties; 31 August 1998.
3. Buckhorn Island; 640 acres; Erie County; 31 August 1998.
4. Iona Island/Doodletown; 1,500 acres; Rockland County; 31 August 1998.
5. Catskill High Peaks; 3,700 acres; Greene and Ulster Counties; 10 June 1999.
6. Nissequogue River; 153 acres; Suffolk County; 28 April 2000.
7. Montezuma Wetlands Complex; 6,449 acres; Seneca, Wayne, and Cayuga Counties; 5 May 2000.
8. Braddock Bay; 2,576 acres; Monroe County; 5 May 2000.
9. Mongaup Valley; 11,967 acres; Sullivan County; 16 June 2000.
10. Bashakill; 2,213 acres; Sullivan County; 16 June 2000.
11. Fahnestock State Park—Hubbard Perkins Conservation Area; 10,050 acres; Putnam County; 29 September 2000.
12. Constitution Marsh; 270 acres; Putnam County; 18 May 2001.
13. Sterling Forest®; 16,833 acres; Orange County; 26 October 2001.
14. Harbor Herons; 111 acres; Richmond County; 17 November 2001.
15. Perch River; 7,862 acres; Jefferson County; 17 November 2001.
16. Adirondack Sub-alpine Forest; 69,000 acres; Essex, Franklin, Hamilton Counties; 17 November 2001.
17. Champlain Marshes; 2,800 acres; Clinton, Essex, and Washington Counties; 9 March 2002.
18. High Tor; 6,100 acres; Ontario and Yates Counties; 12 March 2002.
19. Schodack Island; 864 acres; Rensselaer, Columbia, and Greene Counties; 19 June 2002.
20. Carters Pond; 447 acres; Washington County; 22 October 2002.
21. Oak Orchard/Tonawanda; 8,116 acres; Niagara, Orleans, and Genesee Counties; 22 October 2002.
22. Pharsalia; 10,000 acres; Chenango County; 22 October 2002.
23. Upper and Lower Lakes; 8,781 acres; St. Lawrence County; 22 October 2002.
24. Ashland; 2,037 acres; Jefferson County; 6 May 2003.
25. Long Pond; 394 acres; Chenango County; 6 May 2003.
26. Helderberg; 6,716 acres; Albany County; 6 February 2004.
27. South Shore Tidal Wetlands; 1,377 acres; Nassau and Suffolk Counties; 6 February 2004.
28. John Boyd Thacher/Thompson's Lake; 1,800 acres; Albany County; 27 April 2004.

Literature Cited

Brown, S., C. Hickey, and B. Harrington, Eds. 2000. *The U.S. Shorebird Conservation Plan*. Manomet Center for Conservation Sciences, Manomet, MA.

Kushlan, J.A., M.J. Steinkamp, K.C. Parsons, J. Capp, M. Acosta Cruz, M. Coulter, I. Davidson, L. Dickson, N. Edelson, R. Elliot, R.M. Erwin, S. Hatch, S. Kress, R. Milko, S. Miller, K. Mills, R. Paul, R. Phillips, J.E. Saliva, B. Sydeman, J. Trapp, J. Wheeler, and K. Wohl. 2002. *Waterbird Conservation for the Americas: The North American Waterbird Conservation Plan, Version 1.* Waterbird Conservation for the Americas, Washington, D.C., U.S.A.

North American Waterfowl Management Plan Committee. 1999. *Expanding the Vision: 1998 Update, North American Waterfowl Management Plan.* U.S. Fish and Wildlife Service, Washington, D.C. USA.

Rich, T.D., C.J. Beardmore, H. Berlanga, P.J. Blancher, M.S.W. Bradstreet, G.S. Butcher, D. Demarest, E.H. Dunn, W.C. Hunter, E. Inigo-Elias, J.A. Kennedy, A. Martell, A. Panjabi, D.N. Pashley, K.V. Rosenberg, C. Rustay, S. Wendt and T. Will. 2003. *Partners in Flight North American Landbird Conservation Plan.* Cornell Lab of Ornithology. Ithaca, NY.

U.S. NABCI Committee. 2000. *North American Bird Conservation Regions: Bird Conservation Region Descriptions.* North American Bird Conservation Initiative, U.S. Fish and Wildlife Service, Arlington, VA. [Online version available at: http://www.nabci-us.org/bcrs.html].

Wells, J.V. 1998. *Important Bird Areas in New York State.* National Audubon Society of New York State. Albany, NY.

Chapter 2
Site Identification and Criteria

The identification of IBAs at its most basic level is about determining the most important places for bird species vulnerable to habitat loss or disturbances. Examples include species that are rare or threatened, congregate in large numbers in one place at one time, or are restricted in distribution or to a particular habitat or region.

The Global IBA Approach

BirdLife International developed a set of global criteria to be used by their partner organizations in a rigorous and consistent process to identify IBAs. These standardized, scientific criteria provide both a conceptual framework and quality assurance regarding site identification, guaranteeing that the most important sites will be identified for the right species in the appropriate places. The four global IBA criteria, which form the foundation for the New York IBA criteria, are summarized below.

Global Criterion 1: Sites for Threatened Species

Under this criterion, sites are identified for those species most threatened with extinction at the global level, including species classified as Critical, Endangered, and Vulnerable according to the International Union for Conservation of Nature and Natural Resources Red List Programme (IUCN/SSC 1994) and *Threatened Birds of the World* (BirdLife International 2000). In addition, IBAs are identified for species classified as Conservation Dependent, Data Deficient, or Near Threatened, though they are not globally threatened.

Global Criterion 2: Sites for Range-restricted Species

Under this criterion, IBAs are identified for endemic species with severely restricted ranges of less than 50,000 km^2, within both "Endemic Bird Areas" and "Secondary Areas." An Endemic Bird Area is defined as a region to which two or more restricted-range species are confined, and a Secondary Area is defined as the location where one range-restricted species occurs (Stattersfield *et al.* 1998). Note that this criterion is not used in New York, because the only range-restricted bird in the state, Bicknell's Thrush, is adequately covered under New York's Sites for Species at Risk criterion (defined below).

Global Criterion 3: Sites for Biome-restricted Species Assemblages

Under this criterion, IBAs are identified if a site holds a significant component of a group of species whose distributions are largely or wholly confined to one global biome. This applies to groups of species with distributions of greater than 50,000 km^2 that occur mostly or wholly within a particular biome and are of global importance (Fishpool and Evans 2001). These species may be common within the planning area (e.g., a country), but not widespread outside of it; the planning area has responsibility for their long-term conservation. Many of these assemblages are found in large areas of relatively intact and contiguous habitat. In the U.S., "avifaunal biomes" (defined in the Partners in Flight North American Landbird Conservation Plan; see Rich *et al.* 2003) are utilized in place of global biomes, and Bird Conservation Regions (BCRs, ecological units derived for the purpose of planning and implementing bird conservation in North America; see U.S. NABCI Committee 2000) are used when applying this criterion at the continental level.

Global Criterion 4: Sites for Congregatory Species

Under this criterion, sites are identified if they regularly support significant numbers of birds that are vulnerable as a consequence of their congregatory behavior, either at breeding colonies or during the non-breeding season. In the U.S., a site may qualify under this criterion if it meets any one or more of the four subcriteria listed below:

i The site regularly supports at least 1% of the North American population of a congregatory waterbird species simultaneously, or 5% over a season.

ii The site regularly supports at least 1% of the global population of a congregatory seabird or terrestrial bird species simultaneously, or 5% over a season.

iii The site regularly supports at least 20,000 waterbirds, or at least 10,000 pairs of seabirds, of one or more species.

iv The site is a 'bottleneck site' where at least 5% of the North American population of a migratory waterbird species, or at least 5% of the global population of a migratory seabird or terrestrial species, passes regularly during spring or autumn migrations.

Identification of Global and Continental IBAs in the U.S.

Only the National IBA Technical Committee convened by Audubon has the authority to identify global and continental IBAs in the U.S.. However, to assure that the IBA program remains a grassroots-driven process, all sites evaluated by the National Technical Committee must first be recognized by a state IBA program. Once IBAs have been identified by a state IBA program, the state coordinator or technical committee determines which sites merit recognition at the continental or global level and nominates them to the national IBA staff. The National Technical Committee then reviews the site data and determines if there is justification for recognizing the sites as global or continental level IBAs. Sites that are accepted are publicized as such. Sites that are rejected continue to be recognized as state level IBAs. In other words, decisions of the National Technical Committee do not overrule decisions of the state technical committee with regard to the state-level status of an IBA.

The New York State IBA Approach

The identification of IBAs in New York State follows the formula established by BirdLife. That is, a technical committee evaluates nominated sites relative to the standard criteria to consistently and rigorously determine which sites merit identification as IBAs. During the second round of identifying IBAs in New York, a 24-member IBA Technical Committee (hereafter "committee") was convened to provide advice and oversee the process. The committee was composed of professional conservationists and ornithologists from state and federal agencies and non-profit conservation organizations, and regional bird experts from across New York (see table 2.1). Throughout the site identification process, professional staff of Audubon New York collaborated with the committee to interpret and refine the IBA criteria, develop approaches for each of the criteria, review potential IBAs, and ultimately make decisions about which sites would be identified as IBAs.

New York's original IBA criteria were developed with input from the first IBA Technical Committee in 1996 (see Wells 1998). The current versions of the criteria (below) have been modified slightly from the originals in order to more fully align them with the global IBA criteria developed by BirdLife. A site meeting any of the criteria or subcriteria may qualify as an IBA. Many sites listed in Chapter 3 met more than one criterion.

Northern Harrier

Table 2.1 Members of the New York IBA Technical Committee, which oversaw the second round of site identification

Name	*Affiliation*
Elliott Adams	Central New York Region
Oakes Ames	Audubon New York Board of Directors; New York City Region
Tim Baird	Past President, New York State Ornithological Association (NYSOA); Western New York Region
Bob Budliger	Audubon Society of the Capital Region; New York Breeding Bird Atlas Steering Committee
Michael Burger	Director of Bird Conservation, Audubon New York; Member, National IBA Technical Committee
Valerie Freer	Chair, New York Breeding Bird Atlas Steering Committee; Sullivan County Audubon Society
John Fritz	Long Island Region
Peter Gibbs	St. Lawrence Valley Program Biologist, Ducks Unlimited
Jane Graves	New York Breeding Bird Atlas Regional Co-Coordinator
Jay Greenberg	Rochester Birding Association; Genesee Valley Audubon Society
Lee Harper	Environmental Consultant; Northern New York Region
Sheila Hess	Regional Biologist, Ducks Unlimited
Eric Lind	Director, Constitution Marsh Sanctuary and Audubon Center; Hudson Valley Region
Jillian Liner	IBA Program Coordinator, Audubon New York
Kevin McGowan	Cornell Lab of Ornithology; President, NYSOA
Marty McGuire	New York State Office of Parks, Recreation & Historic Preservation (NYS OPRHP), Taconic Region
Ralph Odell	NYS OPRHP, Taconic Region
Bill Ostrander	NYSOA Kingbird Regional Editor; Chemung Valley Audubon Society
Ray Perry	Bird Conservation Area Program Coordinator, NYS OPRHP
Mike Richmond	Leader, United States Fish & Wildlife Service Cooperative Research Unit; Cornell University
Ken Rosenberg	Director of Conservation Science, Cornell Lab of Ornithology; Co-chair, Partners In Flight International Science Committee; Representative, U.S. National PIF Council
Seymour (Sy) Schiff	Long Island Region
Bryan Swift	Leader, Nongame and Habitat Unit, New York State Department of Environmental Conservation
David VanLuven	Executive Director, New York Natural Heritage Program

New York IBA Criteria

Criterion NY-1: Sites for Species at Risk

Definition

This criterion addresses sites that support significant populations of species at risk. Under this criterion, sites are identified that support a significant breeding, wintering, or migrating population of a species that is listed in New York State as endangered, threatened, or of special concern; a federally listed species; or a species on the Audubon WatchList 2002 (National Audubon Society 2002; see Table 2.2 for a list of species at risk). Sites should support the species with some regularity. Sites where species at risk occur only in overhead flight (e.g., raptor migration bottlenecks) are not included under this criterion. Sites with historic significance and future potential for species at risk are considered. Sites are also considered if they support smaller populations of many species at risk.

Approach

This criterion is based on the global and continental "threatened species" IBA criteria. It has been stepped-down to the state level by expanding the list of species to include species of continental concern and species listed as endangered, threatened, or of special concern in New York State. Additionally, site-level thresholds that are lower than the corresponding global and continental thresholds were established for each species based on at-risk categories, dispersion pattern, and taxonomic group, following the approach used in other countries by BirdLife and adopted by Audubon's National IBA Technical Committee (see Appendix A for New York's site thesholds). To achieve this, the committee grouped species at risk into three categories: Severely at Risk, Highly at Risk, and At Risk. Species were assigned to these categories based upon their threatened status and their population and distribution in New York. A fourth category was developed for "geographically peripheral" species. These species are extremely uncommon or occur so infrequently that manageable populations do not occur in New York; their conservation cannot be adequately addressed through the IBA program. See Box 2.1 for details of the four categories, and Table 2.2 for a list of the species at risk.

Box 2.1

Categories of Species at Risk Addressed Under Criterion NY-1

At-risk species were categorized to guide the establishment of site-level thresholds. The committee decided that the site identification process should be more inclusive (i.e. identify IBAs that collectively encompass a greater proportion of a species' population) for species that are more threatened, less common, and less widespread; and less inclusive for species that are at risk but relatively more common and more widespread.

Severely at Risk

This category includes species at risk with highly restricted distributions and small populations, excluding geographically peripheral species as described below. This category includes species that are on Audubon's Red WatchList and those breeding species that are on the Amber WatchList or the New York State list of endangered, threatened, and special concern species that have highly restricted distributions in New York (i.e. recorded in fewer than 100 of the 5,335 Breeding Bird Atlas [BBA] blocks that cover the entire state). The goal for these species is to secure a high proportion of their populations collectively within IBAs.

Highly at Risk

This category includes species at risk that have restricted distributions and moderate populations. This category includes Amber WatchList and New York State endangered, threatened, and special concern breeding species that have restricted distributions in New York (i.e. recorded in 100-600 BBA blocks) and non-breeding species on the Amber WatchList. The goal for these species is to secure a moderate proportion of their populations collectively within IBAs.

At Risk

This category includes species at risk that are widely distributed and have relatively large populations compared to other species at risk. This category includes Amber WatchList and New York State endangered, threatened, and special concern species that are widely distributed in New York (i.e. recorded in more than 600 BBA blocks). The goal for these species is to secure a smaller proportion of their populations collectively within IBAs.

At Risk, but Geographically Peripheral

This category includes species at risk that are very uncommon or occur so infrequently that they do not have manageable populations in New York; therefore their conservation cannot be adequately addressed through the IBA program. This category includes breeding species at risk with severely restricted distributions in New York (i.e. recorded in fewer than 10 BBA blocks) and non-breeding species at risk that migrate through the state irregularly. The presence of these species at a site will be noted, but IBAs will not be recognized solely based on their presence.

Table 2.2 New York Species at Risk addressed through Criterion NY-1, arranged by risk category (see Box 2.1) with listed status in parentheses (NY-E, NY-T, and NY-SC denote species listed in New York State as endangered, threatened, and of special concern, respectively; WL-R and WL-A denote species on the Red and Amber Audubon WatchLists, respectively).

Severely At Risk

Spruce Grouse (NY-E)
Bald Eagle (NY-T, federally threatened in lower 48 states)
Peregrine Falcon (NY-E)
Piping Plover (NY-E, federally endangered in Great Lakes, federally threatened along the Atlantic Coast, WL-R)
American Oystercatcher (WL-A)
Roseate Tern (NY-E, federally endangered along Atlantic Coast south to NC)
Least Tern (NY-T)
Black Tern (NY-E)
Black Skimmer (NY-SC)
Short-eared Owl (NY-E, WL-A)
Sedge Wren (NY-T)
Bicknell's Thrush (NY-SC, WL-R)
Golden-winged Warbler (NY-SC, WL-R)
Bay-breasted Warbler (WL-A)
Cerulean Warbler (NY-SC, WL-R)
Yellow-breasted Chat (NY-SC)
Henslow's Sparrow (NY-T, WL-R)
Saltmarsh Sharp-tailed Sparrow (WL-R)
Seaside Sparrow (NY-SC, WL-A)

Highly At Risk

Brant (WL-A)
Common Loon (NY-SC)
Pied-billed Grebe (NY-T)
American Bittern (NY-SC)
Least Bittern (NY-T)
Osprey (NY-SC)
Cooper's Hawk (NY-SC)
Northern Goshawk (NY-SC)
Upland Sandpiper (NY-T)
Red Knot (WL-A)
Purple Sandpiper (WL-A)
Short-billed Dowitcher (WL-A)
Common Tern (NY-T)
Common Nighthawk (NY-SC)
Whip-poor-will (NY-SC)
Red-headed Woodpecker (NY-SC, WL-A)
Olive-sided Flycatcher (WL)
Worm-eating Warbler (WL-A)
Canada Warbler (WL-A)
Grasshopper Sparrow (NY-SC)
Rusty Blackbird (WL-A)

At Risk

American Black Duck (WL-A)
Northern Harrier (NY-T)
Sharp-shinned Hawk (NY-SC)
Red-shouldered Hawk (NY-SC)
American Woodcock (WL-A)
Willow Flycatcher (WL-A)
Horned Lark (NY-SC)
Wood Thrush (WL-A)
Blue-winged Warbler (WL-A)
Prairie Warbler (WL-A)
Vesper Sparrow (NY-SC)

At Risk, but Geographically Peripheral

Golden Eagle (NY-E)
Black Rail (NY-E, WL-R)
King Rail (NY-T)
American Golden-Plover (WL-A)
Eskimo Curlew (NY-E, extirpated, WL-R)
Whimbrel (WL-A)
Hudsonian Godwit (WL-A)
Marbled Godwit (WL-A)
Buff-breasted Sandpiper (WL-R)
Wilson's Phalarope (WL-A)
Loggerhead Shrike (NY-E)
Prothonotary Warbler (WL-A)
Kentucky Warbler (WL-A)
Dickcissel (WL-A)

Criterion NY-2:
Sites for Responsibility Species Assemblages

Definition

This criterion identifies sites with the most important habitats for assemblages of bird species whose long-term conservation is the responsibility of New York State. Sites meeting this criterion usually consist of large, intact areas that support all or most of the responsibility species in any one habitat-species assemblage. However, small sites containing exceptional remnants of a habitat type are also considered.

Black-throated Blue Warbler

Approach

This criterion is based on the global and continental "biome-restricted species assemblages" IBA criteria. As at the global and continental levels, this criterion pertains to sites that support groups of species that are restricted to certain regions and habitats. Although these species may be common within the planning area, they are not, by definition, widespread outside of it. The planning area, therefore, bears responsibility for their long-term conservation, even if they are not currently declining or threatened. For application at the state level, this criterion was stepped-down by expanding the list of species to which it applies to include breeding species of high regional responsibility within the BCRs that make up New York State (see Figure 2.1), as determined by applying the PIF species assessment process (Carter *et al.* 2000) to birds of all taxa in the Rocky Mountain Bird Observatory PIF database (RMBO 2002).

Responsibility species were grouped into habitat assemblages for each BCR (see Table 2.3). Because many responsibility species are common in New York, not every site that supports them is identified as a potential IBA. Rather, a "reserve design" approach was taken, targeting the most important 10% of habitat for each assemblage within each BCR as potential IBAs. We defined "most important" as sites with the largest, most intact (e.g., least fragmented) patches of habitat, supporting the highest number of responsibility species composing each assemblage, and with the greatest chance of long-term protection. The 10% target was chosen because it is often suggested as a minimum area for a representative nature reserve network (e.g., IUCN 1980, Noss 1996). Within BCRs, habitat selection was stratified across New York's ecoregions (see map in Edinger *et al.* 2002) to include important ecological variation, and to prevent IBAs from being clustered in only one part of a BCR (Anderson *et al.* 1999).

A Geographic Information System (GIS, using ArcView 3.2, ESRI 1999) spatial analysis of habitat and bird distribution data was undertaken to identify potential IBAs for responsibility species-habitat assemblages. The analysis focused on habitat for forest, shrub, and grassland species assemblages in particular regions of the state. Wetlands in the Atlantic Northern Forest BCR and maritime marshes and beach/dune habitats in the New England/Mid-Atlantic Coast BCR were not addressed in this analysis because those habitats and species are covered adequately by other IBA criteria and existing IBAs. Land cover, predicted breeding bird distribution, and land stewardship data from the New York Gap Analysis Program (NYCFWRU 1998) were used in the analysis. Ancillary data included major and minor roads from the New York State Department of Transportation.

The spatial analysis was conducted in two parts for each habitat/BCR combination. On a landscape scale, the character of habitat and presence of birds on lands unfragmented by major roads was quantified. Landscape context and habitat contiguity were central to the analysis due to their importance to self-sustaining populations of birds as reported in the avian conservation literature (Donovan *et al.* 1997, Faaborg 2002). The highest ranked habitat blocks at the landscape scale were analyzed at a finer, patch level. At the patch scale, the largest, most contiguous, and least fragmented habitats with the greatest chance of long-term protection were determined. These areas were considered potential IBAs and targeted for subsequent verification of species presence.

Presence of predicted responsibility bird species at potential IBAs was verified with preliminary data from the New York Breeding Bird Atlas 2000 (BBA) or through field surveys conducted by Audubon staff and qualified volunteers. The BBA provided recent (2000-2003) data regarding the presence of breeding birds in 5km x 5km blocks across the state. Use of these data was restricted to BBA blocks nearly completely encompassed within potential IBAs. IBAs not adequately covered by the preliminary BBA data were targeted for field surveys during the 2003 breeding season. Habitat quality was also assessed during these surveys.

Table 2.3. New York "Responsibility Species" addressed through Criterion NY-2 by Bird Conservation Region and habitat type. Responsibility species excluded from spatial analysis due to unique habitat requirements or limited distribution, but noted where present, are indicated by an *.

Lower Great Lakes/St. Lawrence Plain Bird Conservation Region (BCR 13)

Forest
Sharp-shinned Hawk
Black-billed Cuckoo
Eastern Wood-Pewee
Wood Thrush
Cerulean Warbler*
Rose-breasted Grosbeak
Baltimore Oriole

Grassland
Killdeer
Upland Sandpiper*
Henslow's Sparrow*
Bobolink
Eastern Meadowlark

Shrub-scrub
American Woodcock
Willow Flycatcher
Brown Thrasher
Blue-winged Warbler
Golden-winged Warbler*
Eastern Towhee
Field Sparrow

Atlantic Northern Forest Bird Conservation Region (BCR 14)

Forest
Ruffed Grouse
American Woodcock*
Black-billed Cuckoo
Yellow-bellied Sapsucker
Eastern Wood-Pewee
Least Flycatcher
Great Crested Flycatcher
Blue-headed Vireo
Veery
Bicknell's Thrush*
Wood Thrush
Northern Parula
Chestnut-sided Warbler
Black-throated Blue Warbler
Black-throated Green Warbler
Blackburnian Warbler
Blackpoll Warbler*
Black-and-white Warbler
American Redstart
Ovenbird
Canada Warbler
Scarlet Tanager
Rose-breasted Grosbeak
Purple Finch*

Wetland
American Black Duck
Hooded Merganser
American Bittern

Appalachian Mountains Bird Conservation Region (BCR 28)

Forest

Sharp-shinned Hawk
Black-billed Cuckoo
Northern Flicker
Eastern Wood-Pewee
Least Flycatcher
Yellow-throated Vireo
Blue-gray Gnatcatcher
Wood Thrush
Black-throated Blue Warbler
Cerulean Warbler*
Black-and-white Warbler
Worm-eating Warbler
Louisiana Waterthrush
Hooded Warbler
Canada Warbler
Scarlet Tanager
Rose-breasted Grosbeak

Grassland

American Kestrel
Henslow's Sparrow*
Bobolink

Shrub-scrub

American Woodcock
Whip-poor-will*
Gray Catbird
Brown Thrasher
Blue-winged Warbler
Golden-winged Warbler
Prairie Warbler
Eastern Towhee
Field Sparrow
Indigo Bunting

Belted Kingfisher*

New England/Mid-Atlantic Coast Bird Conservation Region (BCR 30)

Forest

Broad-winged Hawk
Black-billed Cuckoo
Hairy Woodpecker
Northern Flicker
Eastern Wood-Pewee
Great Crested Flycatcher
Yellow-throated Vireo*
Wood Thrush
Black-and-white Warbler
Worm-eating Warbler*
Louisiana Waterthrush*
Scarlet Tanager
Rose-breasted Grosbeak
Baltimore Oriole

Wetland

American Black Duck
Glossy Ibis
Clapper Rail
Virginia Rail
Marsh Wren
Saltmarsh Sharp-tailed Sparrow
Seaside Sparrow

Shrub-scrub

Northern Bobwhite
American Woodcock
Whip-poor-will
Eastern Kingbird
Gray Catbird
Brown Thrasher
Blue-winged Warbler
Prairie Warbler
Eastern Towhee
Field Sparrow

Beach/Dune

Piping Plover
American Oystercatcher
Common Tern
Least Tern
Black Skimmer

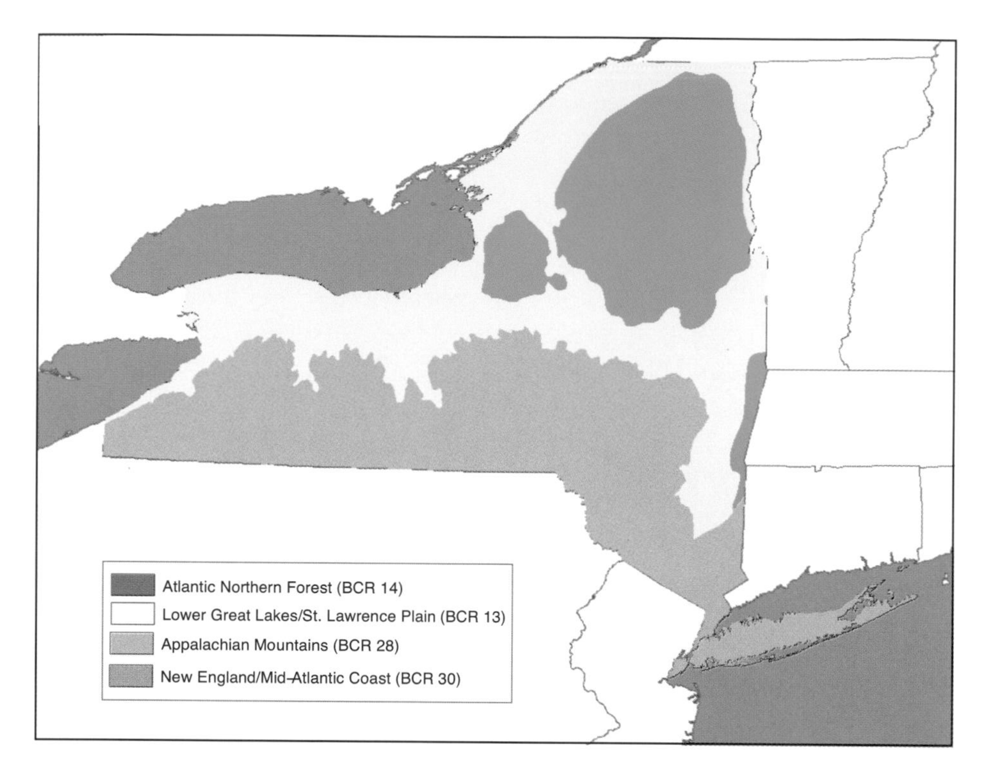

Figure 2.1. The four Bird Conservation Regions (BCRs) that make up New York State.

Waterfowl at Montezuma Wetlands Complex

Criterion NY-3: Sites for Congregations of Birds

Definition

This criterion addresses sites where birds congregate in significant numbers, such as dense populations of breeding birds (e.g., heronries), large numbers of waterfowl or shorebirds in any season, and migratory "bottlenecks" where geographical features such as ridges or shorelines concentrate large numbers of migratory birds. The numerical subcriteria (3a-3e) are guidelines only, and other factors (quality and location of habitat, distribution and importance of species, etc.) are considered. Subcriterion 3a excludes sedentary Canada Geese and Mallards.

Subcriteria:

(3a) The site regularly supports at least 2,000 waterfowl (at one time) during some part of the year. The designation "waterfowl" includes such birds as loons, grebes, cormorants, geese, ducks, coots, and moorhens.

(3b) The site regularly supports at least 100 pelagic seabirds and/or terns (at one time), or 10,000 gulls (at one time) during some part of the year. Human-made food sources for gulls (landfills, dumpsites, sewage outflows, etc.) will not be considered as IBAs. The designation "pelagic seabird" includes such birds as shearwaters, storm-petrels, fulmars, gannets, jaegers, and alcids.

(3c) The site regularly supports at least 300 shorebirds (at one time) if an inland site, or 1,000 shorebirds (at one time) if coastal, during some part of the year. The designation "shorebirds" includes such birds as plovers, sandpipers, snipe, woodcock, and phalaropes.

(3d) The site regularly supports at least 100 wading birds during some part of the year. The designation "wading birds" includes such birds as bitterns, herons, egrets, and ibises.

(3e) The site regularly serves as a migratory corridor or "bottleneck" for at least 8,000 raptors (seasonal total) during spring or fall migration.

(3f) The site supports an exceptional number and/or diversity of landbirds during migration. This includes sites that provide important habitat for more landbirds than are found at most other sites during migration. No absolute thresholds have been set due to the scarcity of quantitative data. Sites should be clearly unique from other sites in the local area.

(3g) The site supports a significant number of individuals of a particular species, but may support a smaller total number of birds than any of the subcriteria above (3a-3f). Ideally, the site should hold more than 1% of the state (if breeding) or flyway (if migrating) population of a species.

Approach

This criterion is based on the global and continental "congregatory species" IBA criteria, which were stepped-down to the state level by establishing lower site-level thresholds than those used at the global and continental levels.

Second Round Site Review Process

In concept, the identification of IBAs is a simple process. Sites are nominated for consideration under one or more of the criteria, the committee reviews the data, and sites meeting at least one criterion are accepted as IBAs. In reality, the process was more complex and iterative.

The second round of identifying IBAs began in the fall of 2002 when the nomination process was officially opened and nomination packets were sent to local Audubon chapters, bird clubs, conservation professionals, and individuals who nominated sites during the first round in 1996. Early in 2003, the committee met to discuss and plan the process they would use for reviewing and approving IBAs. Several decisions made at that time significantly influenced the scope and direction of the second round, including the decision that all existing IBAs would be re-assessed and that the three IBA criteria would be tackled sequentially, such that all sites would be reviewed with regard to one criterion, and decisions would be made before proceeding to the next criterion.

As the committee reviewed sites under each criterion, significant efforts were made to secure additional data for any sites lacking sufficient data for a thorough evaluation. Additional data were sought by contacting the people who had nominated the sites (both existing IBAs and newly-nominated sites) as well as others familiar with the sites, and by scouring existing databases for pertinent data. Sources of data for this latter effort included the New York Natural Heritage Program database, International Shorebird Surveys, Winter Waterfowl Surveys, Long Island Colonial Waterbird Surveys, Christmas Bird Counts, New York Breeding Bird Atlas, and eBird. In some cases, data from these existing sources led to the identification of new potential IBAs. For all criteria, only data from the past 10 years were accepted in support of site nominations, in keeping with the policy of Audubon's national IBA program.

Next, all of the supporting data were entered into a central IBA database from which site summaries were compiled and sent to regional subgroups of the committee for review. Initial acceptance and rejection decisions and comments were collected from each member of the regional subgroups and organized. Conference calls of the regional subgroups were then held, and consensus decisions of acceptance or rejection were made for most sites. For some sites additional information was requested and obtained before a decision was made.

After running through this general process for all three criteria, a total of 172 potential IBAs had been reviewed by the committee. Many potential IBA sites were reviewed under more than one criterion: 140 sites were reviewed under the Species at Risk criterion, 107 sites under the Responsibility Species Assemblage criterion, and 104 sites under the Congregations criterion.

Finally, the full committee met in the fall of 2004 to make a final, complete review of the decisions reached by the subgroups regarding all sites and criteria. In addition, the de-listing of previously identified IBAs that did not qualify during the second round (see below) and the designation of Important Bird Research Areas (see Wells 1998) were discussed. The committee decided that research areas should not be included in this book, because no equivalent global or continental criteria exist, objective criteria had not been established, and a comprehensive review of potential "research" sites had not been made. In the final analysis, 136 sites were approved as IBAs. Descriptions and data summaries for those sites appear in Chapter 3.

De-listing IBAs

In order to maintain the integrity of the IBA designation, the process of identifying IBAs must remain consistent, science-based, and within the conceptual framework supplied by the global criteria. The result is a global network of the most critical sites for birds and bird conservation. Sites that do not merit IBA status may still be important, either at the local level or as part of the vast matrix of habitats that helps support birds as they complete their annual cycles. In identifying certain sites as IBAs, Audubon New York does not intend to diminish efforts to conserve or monitor other sites. In fact, we encourage protection and proper management of all habitats throughout the state.

Sometimes, sites may cease to qualify as IBAs due to deterioration of the habitat quality to the point that restoration is not an option, persistent declines of species that trigger IBA criteria to levels below site thresholds, or re-interpretation of IBA criteria or revision of thresholds as a result of new information. For these reasons, the committee developed a process for de-listing IBAs. The process requires several years of recent data demonstrating that a site no longer meets any of the criteria thresholds as well as a consensus opinion that the site no longer has the capability of achieving IBA status. A lack of data supporting the continued IBA status of a site is not sufficient grounds for de-listing it.

As part of the identification of IBAs for this book, the committee determined that the conditions to de-list a site had been met for six sites previously identified as IBAs (in Wells 1998), and that it was necessary to de-list them. Those six sites are Carter Pond, Ferd's Bog, Franklin Mountain Hawkwatch, Gilboa Reservoir, Ticonderoga Marsh, and Webb Royce Swamp. Though de-listing these sites as IBAs was necessary for programmatic consistency, doing so does not lessen their ecological or recreational value, or their significance as open space.

Literature Cited

Anderson, M.G., P. Comer, D. Grossman, C. Groves, K. Poiani, M. Reid, R. Schneider, B. Vickery, and A. Weakley. 1999. *Guidelines for representing ecological communities in ecoregional conservation plans.* The Nature Conservancy.

BirdLife International. 2000. *Threatened birds of the world.* Barcelona and Cambridge, UK: Lynx Edicions and BirdLife International.

Carter, M.F., W.C. Hunter, D.N. Pashley, and K.V. Rosenberg. 2000. Setting conservation priorities for landbirds in the United States: the Partners in Flight approach. Auk 117: 541-548.

Donovan, T. M., P. W. Jones, E. M. Annand, and F. R. Thompson III. 1997. Variation in local-scale edge effects: mechanisms and landscape context. Ecology 78: 2064-2075.

Edinger, G. J., D. J. Evans, S. Gebauer, T. G. Howard, D. M. Hunt, A M. Olivero (editors). 2002. *Ecological Communities of New York State. Second Edition. A revised and expanded edition of Carol Reschke's Ecological Communities of New York State.* New York Natural Heritage Program, New York State Department of Environmental Conservation, Albany, NY.

ESRI. 1999. Environmental Systems Research Institute. Redlands, CA.

Faaborg, J. R. 2002. *Saving Migrant Birds: Developing Strategies for the Future.* University of Texas Press. Austin, TX.

Fishpool, L. D. C. and M. I. Evans, eds. 2001. *Important Bird Areas in Africa and associated islands: Priority sites for conservation.* Newbury and Cambridge, UK: Pisces Publications and BirdLife International (BirdLife Conservation Series No. 11).

International Union for Conservation of Nature and Natural Resources (IUCN). 1980. *World conservation strategy: living resource conservation for sustainable development.* IUCN-UNEP-WWF. Gland, Switzerland.

IUCN/SSC 1994. *IUCN Red Data List categories.* Gland, Switzerland: IUCN Species Survival Commission.

National Audubon Society. 2002. Audubon WatchList 2002: An early warning system for bird conservation. [Online version available at: http://www.audubon.org/birds/watchlist/index.html].

Noss, R. F. 1996. in *National Parks and Protected Areas: Their Role in Environmental Protection,* R. G. Wright, Ed. (Blackwell Science, Cambridge, MA, 1996), pp. 91-119.

NYCFWRU (New York Cooperative Fish and Wildlife Research Unit and Cornell Institute for Resource Information Systems). 1998. The New York Gap Analysis Project Home Page, Version 98.05.01. Departments of Natural Resources and Soil, Crop, and Atmospheric Sciences, Cornell University, Ithaca, NY.

Rocky Mountain Bird Observatory. 2002. Partners in Flight Database. http: //www.rmbo.org/pif/pifdb.html.

Stattersfield, A.J., M.J. Crosby, A.J. Long and D.C. Wege. 1998. *Endemic bird areas of the world: Priorities for biodiversity conservation.* Cambridge, UK: BirdLife International (BirdLife Conservation Series 7).

Wells, J.V. 1998. *Important Bird Areas in New York State.* National Audubon Society of New York State. Albany, NY.

Chapter 3
IBA Site Summaries

The IBA sites have been organized geographically, with the state divided into four regions: Western New York, Northern New York, The Catskills and Hudson Valley, and New York City and Long Island (a map showing the four regions is found at the front of the book). Within each region, the sites are listed alphabetically by name. A map at the beginning of each section shows the IBA locations within each region and associated page numbers where the site descriptions can be found. A list of the sites and corresponding page numbers can also be found in the index at the back of the book.

How to Read the Site Summaries

Site summaries include a general site description, the IBA criteria met and corresponding supporting data, additional bird information, and relevant conservation information. The summaries were compiled from information presented on the nomination forms, published and unpublished literature, and personal communication with individuals familiar with the areas.

Site summary sample:

Site Name

Site names usually are based on a local landmark, feature, or already named parcel of land (e.g. a state park).

Municipality and Counties

Indicates the towns and counties in which the site occurs. If a site occurs in more than three towns or four counties, the site summaries will read "Multiple municipalities" or "Multiple counties."

This icon indicates that the site is a suitable location for birding. That is, the site has public access, an obvious location from which to view birds, and in some instances, infrastructure for visitors. (Icon courtesy of Watchable Wildlife. This binocular logo is an international road sign directing you to a quality wildlife viewing site. For more information on this program visit www.watchablewildlife.org)

Size
The approximate size of the site in acres.

Latitude
The latitude of the approximate center of the site in decimal degrees.

Elevation
The approximate minimum and maximum elevation of the site in feet.

Longitude
The longitude of the approximate center of the site in decimal degrees.

IBA Criteria Met

Criterion	*Species*	*Data*	*Season*	*Source*
Criterion met	Species triggering the criterion	Data demonstrating that the criterion was met	Season	Source of data

This table lists the criteria that the site met and corresponding data. Different criteria require different types of data; therefore, the data format may vary by criterion. "Breeds" indicates that the bird has been observed during the breeding season; breeding may or may not have been confirmed. Also, in instances where there is more than one source of data per criterion, a number will indicate each source, e.g. [1] or [2], etc. Sources are listed at the end of the chapter. NY BBA data include the following abbreviations: CO=confirmed breeding, PR=probable breeding, and PO=possible breeding.

Description: A general description of the site, including location, habitats, ownership, and other noteworthy physical or ecological characteristics. If the site met the Responsibility Species Assemblage criterion, the NY GAP "land covers" making up the habitat type are summarized.

The NY GAP land cover classification scheme includes 31 land cover types (Smith *et al.* 2001). It is important to keep in mind that the NY GAP data are based on a coarse mapping scale (1:100,000), are derived from remote sensing, and represent a snapshot in time. Therefore, the data can only be considered a rough estimate of actual land cover. The acreages of GAP land cover included in the site descriptions have not been ground-truthed.

Birds: A summary of the site's importance to birds. Includes information about the species that triggered the IBA criteria and other available bird information. Numbers represent individual birds unless otherwise specified.

Conservation: A summary of conservation activities at the site, including recent land acquisitions, restoration projects, management activities, and monitoring programs. Conservation concerns and needs are also included.

Chapter 3.1
Western New York

Gorge at Letchworth State Park

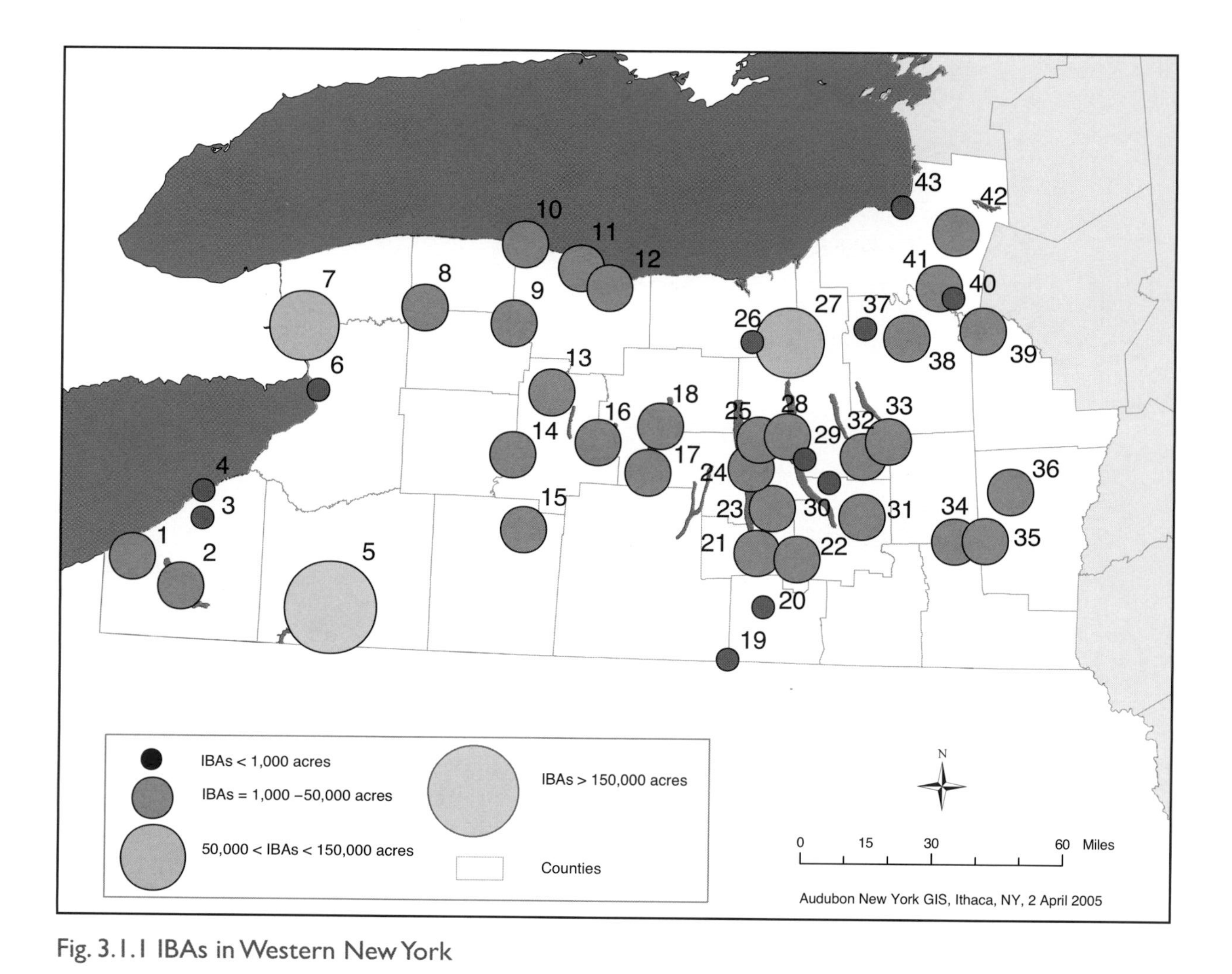

Fig. 3.1.1 IBAs in Western New York

Allegany Forest Tract

Multiple municipalities, Cattaraugus and Chautauqua Counties

195,000 acres
1,310-2,385' elevation

42.1153°N
78.7233°W

IBA Criteria Met

Criterion	*Species*	*Data*	*Season*	*Source*
Species at Risk	Osprey	4CO, 0PR, 2PO	Breeding	NY BBA 2000
Species at Risk	Bald Eagle	4 pairs in 2002-2004; daily counts in winter include at least 25 ind. in 2004, 23 in 2003	Breeding and Winter	Tom LeBlanc pers. comm. 2004
Species at Risk	Sharp-shinned Hawk	2CO, 1PR, 2PO	Breeding	NY BBA 2000
Species at Risk	Cooper's Hawk	Extent of habitat and breeding atlas presence strongly suggest that the threshold is being met [1]; 1CO, 0PR, 4PO [2]	Breeding	1. Technical Committee consensus; 2. NY BBA 2000
Species at Risk	Northern Goshawk	Extent of habitat and breeding atlas presence strongly suggest that the threshold is being met [1]; 2CO, 0PR, 0PO [2]	Breeding	1. Technical Committee consensus; 2. NY BBA 2000
Species at Risk	Red-shouldered Hawk	5+ breeding pairs	Breeding	Tim Baird pers. comm. 2003
Species at Risk	American Woodcock	Extent of habitat and breeding atlas presence strongly suggest that the threshold is being met [1]; 0CO, 2PR, 4PO [2]	Breeding	1. Technical Committee consensus; 2. NY BBA 2000
Species at Risk	Willow Flycatcher	Extent of habitat and breeding atlas presence strongly suggest that the threshold is being met [1]; 0CO, 4PR, 2PO [2]	Breeding	1. Technical Committee consensus; 2. NY BBA 2000
Species at Risk	Wood Thrush	1CO, 10PR, 3PO	Breeding	NY BBA 2000

Criterion	Species	Data	Season	Source
Species at Risk	Blue-winged Warbler	Extent of habitat and breeding atlas presence strongly suggest that the threshold is being met [1]; 2CO, 7PR, 2PO [2]	Breeding	1. Technical Committee consensus; 2. NY BBA 2000
Species at Risk	Cerulean Warbler	57 ind. in 1997, 93 in 1998, 21 in 1999	Breeding	Cerulean Warbler Atlas Project
Species at Risk	Canada Warbler	Extent of habitat and breeding atlas presence strongly suggest that the threshold is being met [1]; 2CO, 4PR, 3PO [2]	Breeding	1. Technical Committee consensus; 2. NY BBA 2000
Responsibility Species Assemblage-Forest	Sharp-shinned Hawk, Black-billed Cuckoo, Northern Flicker, Eastern Wood-Pewee, Least Flycatcher, Yellow-throated Vireo, Blue-gray Gnatcatcher, Wood Thrush, Black-throated Blue Warbler, Cerulean Warbler, Black-and-white Warbler, Louisiana Waterthrush, Hooded Warbler, Canada Warbler, Scarlet Tanager, Rose-breasted Grosbeak	Breed	Breeding	NY BBA 2000
Congregations-Individual Species	Bald Eagle	5% of state Bald Eagle winter population has been documented	Winter	Tim Baird pers. comm. 2003
Congregations-Individual Species	Cerulean Warbler	1%-3% of the estimated state Cerulean Warbler breeding population has been documented	Breeding	Cerulean Warbler Atlas Project

Description: This site includes the Allegany State Park (almost 65,000 acres) and extensive surrounding forested lands. According to the NY GAP land cover data, almost 95% of site is forest habitat, which includes sugar maple mesic, oak, successional hardwood, evergreen northern hardwood, evergreen plantation, and deciduous wetland forests. The Kinzua Dam creates the Allegheny Reservoir, and the reservoir water level varies from year to year.

Birds: This site supports a diverse assemblage of wood warblers (20-plus species) and other representative forest species, including the breeding Sharp-shinned Hawk, Black-billed Cuckoo, Northern Flicker, Eastern Wood-Pewee, Least Flycatcher, Yellow-throated Vireo, Blue-gray Gnatcatcher, Wood Thrush, Black-throated Blue Warbler, Cerulean Warbler, Black-and-white Warbler, Louisiana Waterthrush, Hooded Warbler, Canada Warbler, Scarlet Tanager, and Rose-breasted Grosbeak. At-risk species supported at the site include the American Black Duck (migrant), Common Loon (migrant), Pied-billed Grebe (migrant), American Bittern (migrant), Osprey (breeds), Bald Eagle (breeds), Northern Harrier (migrant), Sharp-shinned Hawk (breeds), Cooper's Hawk (breeds), Northern Goshawk (breeds), Red-shouldered Hawk (breeds), Golden Eagle (rare migrant), American Golden-Plover (rare migrant), Hudsonian Godwit (rare migrant), Short-billed Dowitcher (rare migrant), American Woodcock (breeds), Common Nighthawk (migrant), Olive-sided Flycatcher (migrant),Willow Flycatcher (breeds), Wood Thrush (breeds), Blue-winged Warbler (breeds), Golden-winged Warbler (possible breeder), Bay-breasted Warbler (rare migrant), Cerulean Warbler (breeds), Worm-eating Warbler (rare visitor), Kentucky Warbler (possible breeder), Canada Warbler (breeds), Yellow-breasted Chat (probable breeder), and Rusty Blackbird (migrant). The reservoir is a resting stopover for a diverse group of waterfowl, including large numbers of Tundra Swans under certain weather conditions.

Conservation: A portion of this site is listed in the 2002 State Open Space Conservation Plan as a priority site under the project name Allegany State Park. Long-term protection and stewardship of private lands within the site are needed. Options include public acquisition, purchase of conservation easements, and sustainable forestry agreements. Research conducted at the site includes Red-shouldered Hawk surveys, Saw-whet Owl banding (13 individuals were banded in 2004), and a Monitoring Avian Productivity and Survivorship (MAPS) banding station. Inventory and monitoring, especially of at-risk species, should continue. The forest interior receives a significant amount of trail and other recreational uses. Such uses seem to be compatible with the resources and with bird conservation. The reservoir is heavily used during the summer for recreation. During the first round of IBA site identifications, this site was recognized under the research criterion because long-term research and monitoring projects are based there.

Aurora Grassland Complex

Genoa and Ledyard, Cayuga County

420 acres
750-800' elevation

42.6863°N
76.6505°W

IBA Criteria Met

Criterion	*Species*	*Data*	*Season*	*Source*
Species at Risk	Short-eared Owl	3-7 ind. each year from 1998-2002	Winter	Jeff Wells pers. comm. 2004

Description: This site consists of privately owned agricultural grasslands in Cayuga County on the eastern side of Cayuga Lake.

Birds: Breeding at-risk species found at this site include the Northern Harrier (possible), Upland Sandpiper (documented in the late 1980s and early 1990s), Sedge Wren (territorial individuals in the late 1990s), Grasshopper Sparrow (three in 1998-2003), and Henslow's Sparrow (observed through the late 1990s). Wintering at-risk species include the Northern Harrier (8-15 in 1999-2002), Short-eared Owl (3-7 annually in 1998-2002), and Horned Lark (2-30 in 1999-2002).

Conservation: Threats to this area include vegetational succession and suburban development. Lands in this area are for sale. Some of the land is enrolled in a governmental grassland conservation program, and further participation should be encouraged. Monitoring of at-risk species should be implemented.

Bergen Swamp

Bergen and Byron, Genesee County

3,400 acres
540-630' elevation

43.0995°N
77.9936°W

IBA Criteria Met

Criterion	*Species*	*Data*	*Season*	*Source*
Responsibility Species Assemblage-Forest	Black-billed Cuckoo, Eastern Wood-Pewee, Wood Thrush, Cerulean Warbler, Rose-breasted Grosbeak, Baltimore Oriole	Breeds	Breeding	NY BBA 2000

Description: Bergen Swamp is located between the Niagara and Onondaga limestone escarpments and is owned by a land trust, the Bergen Swamp Preservation Society (BSPS). The swamp is a remnant of the ancient glacial Lake Tonawanda. It includes northern white cedar forest, pine-hemlock forest, and beech-maple deciduous forest. According to the NY GAP land cover data, approximately 90% of the site is forest habitat, which includes deciduous wetland, successional hardwood, and sugar maple mesic forests. The site supports a high diversity of plants, with a total of 2,392 species identified, and is especially known for its orchids. Many of the plants are rare or endangered. The habitat here is unique in New York State and rare nationally. Endangered non-avian fauna are also present. The swamp has been recognized by the U.S. Department of the Interior as a National Natural Landmark.

Birds: This site is rich in breeding birds and has an assemblage of species that do not breed elsewhere in the Lake Ontario plain. These include boreal species such as the Yellow-bellied Sapsucker, Alder Flycatcher, Blue-headed Vireo, Winter Wren, Hermit Thrush, Nashville Warbler, Blackburnian Warbler, Black-and-white Warbler, Canada Warbler, and Purple Finch. It also supports more southern species such as the Acadian Flycatcher. Five species of breeding owls are present: the Great Horned, Eastern Screech, Barred, Long-eared, and Northern Saw-whet Owls. In addition, the following at-risk species are confirmed breeders: the American Bittern, Willow Flycatcher, Wood Thrush, Blue-winged Warbler, and Canada Warbler.

Bergen Swamp

Conservation: Bergen Swamp is a popular place to visit and has the potential to be used for ecotourism. However, BSPS is concerned about overuse. A permit is required for visits by groups of six or more people. BSPS allows research at the swamp, but requires a permit. A limited hunting program in cooperation with surrounding landowners is used to control deer in order to limit browse damage. The effectiveness of these measures is not known. Bergen Swamp is of historic interest as one of the oldest nature preserves to be privately protected by a land trust. It was first chartered by the New York Board of Regents in 1936 as a "Living Museum."

Braddock Bay

Greece and Parma, Monroe County

5,300 acres
240-285' elevation

43.2933°N
77.6951°W

IBA Criteria Met

Criterion	*Species*	*Data*	*Season*	*Source*
Congregations-Raptors	Mixed species	An average of 53,575 and maximum of 144,000 ind. counted during 1993-2004 seasons	Spring	Braddock Bay Raptor Research
Congregations-Waterfowl	Mixed species flocks	Regularly supports waterfowl concentrations in the thousands, Round Pond alone supports 1,000-2,000 ind.	Fall	Nomination, Jay Greenberg
Congregations-Migrant Landbirds	Migrant landbirds	3,000-8,000 ind. of 70-90 species have been banded each spring and fall [1]; thousands of passerines can be found in the woods during migration [2]	Fall and spring	1. Braddock Bay Bird Observatory; 2. Nomination, Jay Greenberg

Description: This site is located on the shore of Lake Ontario near the city of Rochester, and includes ponds, creeks, wetlands, woods, and fields. The wetlands are dominated by cattail marshes while the uplands are predominantly wet deciduous forests, abandoned farmlands, and private residential properties. The site includes the New York State Department of Environmental Conservation (NYS DEC)-administered Braddock Bay Wildlife Management Area, and municipal and private land. Braddock Bay Raptor Research conducts an ongoing hawk banding program and annually staffs a hawkwatch from March through May. Braddock Bay Bird Observatory carries out a large-scale passerine mist-netting and banding operation each fall and spring.

Birds: The Braddock Bay area hosts a remarkable diversity and abundance of birds. The site is well known for having one of the world's largest spring hawk flights (144,000 counted in 1996). Banding efforts have shown the area to be an important owl migration point, with an average of 35 Long-eared Owls and 100 Northern Saw-whet Owls banded each spring from 1985-1995. Woodlands in the area are known to host large numbers and a great variety of songbirds. A passerine banding station at the site has operated annually for the last 19 years and bands thousands of individuals (7,966 individuals banded in 2003). The area also supports breeding populations of at-risk species, including the Pied-billed Grebe, American Bittern, Least Bittern, Northern Harrier, and Sedge Wren. Historically the site has supported nesting Black Terns. The site also regularly hosts thousands of waterfowl during migration.

Conservation: This site is listed in the 2002 State Open Space Conservation Plan as a priority site under the project name Braddock Bay. Portions of this site have been designated as a state Bird Conservation Area. Although much of the wetland marsh habitat is currently protected and under management by the NYS DEC, most of the upland portions have become residential or commercial developments. The remaining forest, shrub, and grassland fragments are vitally important as foraging locations for migrating hawks, owls, and passerines, and are being rapidly lost to development. The NYS DEC and Town of Greece have an ongoing program for acquiring privately owned land in the Braddock Bay area to increase the Wildlife Management Area. Recent purchases include the 68-acre Burger Park (2001) and the 130-acre Dahlheim property (2003), both located on Salmon Creek. The site's most well-known passerine concentration site, Island Cottage Woods, is under threat of development, except for 61 acres transferred to the Genesee Land Trust (GLT) for permanent protection in 1999. The site of the Braddock Bay Bird Observatory's long-term banding operation is also protected by the GLT. In December 2004, GLT acquired five acres near the banding station, adjacent to the West Spit. They, along with the Braddock Bay Bird Observatory and the Rochester Birding Association, serve as managing partners. There have been localized problems with unsupervised ATV use, illegal woodcutting, illegal dumping, and suburban lawn runoff. Die-offs of waterfowl occurred in one area in 1995 and 1996 due to diazinon, a lawn insecticide. The site of the former Odenbach ship building plant on the east side of Round Pond is heavily polluted with organic solvents and heavy metals. Cleanup efforts are ongoing, but the extent of the pollution plume is still unclear.

Canandaigua Lake

Multiple municipalities, Ontario and Yates Counties

10,300 acres
680-760' elevation

42.7821°N
77.3003°W

IBA Criteria Met

Criterion	*Species*	*Data*	*Season*	*Source*
Congregations-Waterfowl	Mixed species flocks	5,535 ind. in 2004, 12,809 in 2003, 13,960 in 2002, 12,567 in 2001, 4,918 in 2000	Winter	NYSOA winter waterfowl counts
Congregations-Individual Species	Mallard	Supported over 1% of estimated 6-year average state winter population (453 ind.) in each of the last 5 years	Winter	NYSOA winter waterfowl counts
Congregations-Individual Species	Redhead	Supported over 1% of estimated 6-year average state winter population (156 ind.) in 4 out of last 5 years, supported over 40% of 2003 pop. and over 15% of 2004 and 2001 pop.	Winter	NYSOA winter waterfowl counts

Description: Canandaigua Lake is one of the smallest of the Finger Lakes. The glacially carved, 15 mile-long, narrow lake spans two counties, with the city of Canandaigua on its north end and Naples on the south. To the west lies Honeoye Lake and to the east, Keuka and Seneca Lakes. Much of the lakeshore is developed or in agriculture, but there are scattered marshes and wetlands. The lake is owned by the state of New York, but the lakeshore includes mostly private, some municipal, and some NYS DEC-administered land.

Birds: This site is an important winter waterfowl area, particularly for Mallards and Redheads. The NYSOA winter waterfowl counts have documented over 1% of the estimated state winter population of Mallards for the last five years, and over 1% of the estimated state winter population of Redheads in four out of the last five years (over 40% of 2003 population and over 15% of 2004 and 2001 populations).

Conservation: The introduction of zebra mussels and non-native fish may have a negative effect on the aquatic ecosystem and the prey base of certain waterfowl species. However, some diving duck species feed extensively on zebra mussels and may benefit from increases in their populations. Manipulation of water levels in the lake may impact waterfowl use of the lake in unknown ways. More research is needed to understand how different species of waterfowl are or will be impacted by changes in the lake ecosystem. The lake is heavily used from spring through fall for recreational boating, fishing, and related activities. Disturbance does not seem to be a major problem since the largest concentrations of waterfowl occur before and after peak boating activity. Pollution from various sources, including agricultural runoff and boats, could impact the aquatic ecosystems that birds rely upon, and should be monitored. Monitoring of waterfowl numbers should continue.

Caswell Road Grassland Complex

Dryden, Tompkins County

1,870 acres
985-1,185' elevation

42.4976°N
76.3923°W

IBA Criteria Met

Criterion	*Species*	*Data*	*Season*	*Source*
Species at Risk	Henslow's Sparrow	5+ breeding pairs in 1995, 10-12 pairs in early 1990s	Breeding	Nomination, Bard Prentiss

Description: This site includes a flat area of abandoned farmland with a high water table and clay soils making it damp much of the year. This area is privately owned, and is leased by a hunt club, which maintains the habitat for Wild Turkey, Ring-necked Pheasant, and white-tailed deer (*Odocoileus virginianus*). Cover consists primarily of grasses and weeds, however, in areas where mowing has ceased, shrub/scrub habitat is encroaching. A small percentage of the site is reverting to white pine, white ash, aspen, and red maple.

Birds: This site has been an important grassland bird nesting area, supporting breeding at-risk species such as Northern Harriers (one pair in 1995), Upland Sandpipers (at least one pair in 1996), Grasshopper Sparrows (at least three pairs in 1995), and Henslow's Sparrows (at least five pairs in 1995, 10-12 pairs in the early 1990s). Other breeding grassland species include Savannah Sparrows and Bobolinks (30 plus pairs in 1995).

Conservation: In recent years the populations of grassland bird species have declined and there is a need for targeted monitoring to better understand the current situation. Management strategies seem compatible with the needs of grassland bird species. Mowing is carried out on a schedule that does not interfere with the nesting season. The land is posted and patrolled, and there is little public misuse. Hunt club members rarely use the site during the breeding season, so there is little disturbance of grassland birds. Development of a long-term grassland management plan in cooperation with the hunt club and the landowner would be highly desirable.

© Jeff Nadler

Savannah Sparrow

Cayuga Lake

Multiple municipalities, Cayuga, Seneca, and Tompkins Counties

42,000 acres
375-435' elevation

42.6873°N
76.7026°W

IBA Criteria Met

Criterion	*Species*	*Data*	*Season*	*Source*
Species at Risk	American Black Duck	1,045 ind. in 2004, 1,296 in 2003, 2,052 in 2002, 1,466 in 2001, 871 in 2000	Winter	NYSOA winter waterfowl counts
Congregations-Waterfowl	Mixed species flocks	41,839 ind. in 2004, 27,401 in 2003, 68, 704 in 2002, 61,259 in 2001, 39,250 in 2000	Winter	NYSOA winter waterfowl counts
Congregations-Individual Species	Canada Geese	Site supported 18% of state estimated winter population in 2000, 37% in 2001, 25% in 2002, 12% in 2003, 23% in 2004	Winter	NYSOA winter waterfowl counts
Congregations-Individual Species	American Black Duck	Site supported 5% of state estimated winter population in 2000, 8% in 2001, 11% in 2002, 10% in 2003, 10% in 2004	Winter	NYSOA winter waterfowl counts
Congregations-Individual Species	Mallard	Site supported 4% of state estimated winter population in 2000, 9% in 2001, 11% in 2002, 3% in 2003, 6% in 2004	Winter	NYSOA winter waterfowl counts
Congregations-Individual Species	Canvasback	Site supported 6% of state estimated winter population in 2000, 12% in 2002, 1% in 2003, 1% in 2003	Winter	NYSOA winter waterfowl counts
Congregations-Individual Species	Redhead	Site supported 30% of state estimated winter population in 2000, 37% in 2001, 39% in 2002, 48% in 2003, 26% in 2004	Winter	NYSOA winter waterfowl counts
Congregations-Individual Species	Common Goldeneye	Site supported 9% of state estimated winter population in 2000, 7% in 2001, 6% in 2002, 4% in 2003, 10% in 2004	Winter	NYSOA winter waterfowl counts

Description: Cayuga Lake is the longest of the Finger Lakes, extending approximately 38 miles from north to south. As with the other Finger Lakes, Cayuga Lake was glacially carved and is relatively narrow. The lake spans three counties, with the City of Ithaca on its southern end and Seneca Falls just west of the northern end. To the west lies Seneca Lake and to the east Owasco Lake. Much of the lakeshore is developed or in agriculture, but there are scattered marshes and wetlands. The lake is owned by State of New York and the lakeshore includes mostly private land with some municipal and state-owned land, including land administered by the New York State Office of Parks, Recreation and Historic Preservation (NYS OPRHP).

Birds: The entire lake is important to waterfowl during migration and winter, supporting high numbers of individuals and a great diversity of species (around 30 species of ducks and geese). During 2000-2004, the lake supported an average of 1,346 American Black Ducks (11% of state wintering population), 2,987 Mallards (8% of state wintering population), 403 Canvasbacks (4% of state wintering population), 6,012 Redheads (45% of state wintering population), 332 scaup, and 993 Common Goldeneyes (8% of state wintering population). Counts of Pied-billed, Horned, and Red-necked Grebes sometimes number in the hundreds during fall and winter. The airspace over the lake is an important pathway for Common Loons migrating from Lake Ontario to Delaware Bay, with count totals of 8,000-10,000 from October through early December. Large numbers of gulls and smaller numbers of terns use the lake, especially during migration, when species include Bonaparte's Gulls (100-1,000 plus), Ring-billed Gulls (one-day counts of 20,000-50,000 plus), and Caspian Terns (30-40 plus).

Conservation: The introduction of zebra mussels and non-native fish may negatively affect the aquatic ecosystem and the prey base of certain waterfowl species. However, some diving duck species feed extensively on zebra mussels and may benefit from their population increase. The lake is heavily used from spring through fall for recreational boating and fishing. Disturbance does not seem to be a major problem since the largest concentrations of waterfowl occur before and after the peak of the boating season. Pollution from various sources, including boats and non-point source agricultural runoff, could impact the aquatic ecosystem and should be monitored. More research is needed to understand how different species of waterfowl are impacted by changes in the lake ecosystem. A Cayuga Lake Watershed Restoration and Protection Plan was written in 2001 under the oversight of the Cayuga Lake Watershed Intermunicipal Organization. The Cayuga Lake Watershed Network is involved in the NYS Citizens Lake Assessment Program, water quality (phosphorus, sediment, and bacteria) monitoring, and invasive plant control.

Chautauqua Lake

Multiple municipalities, Chautauqua County

14,000 acres
1,290-1,470' elevation

42.1703°N
79.4084°W

IBA Criteria Met

Criterion	*Species*	*Data*	*Season*	*Source*
Species at Risk	Pied-billed Grebe	Max. daily fall migration: 20 ind. in 1998; 22 in 1999, 48 in 2000, 120 in 2001, 67 in 2002, 5 in 2003	Fall	Dick Miga pers. comm. 2004
Species at Risk	Common Tern	25 ind. in 1999, 71 in 2000, 1 in 2001, 1 in 2002	Migration	Dick Miga pers. comm. 2004
Congregations-Waterfowl	Tundra Swan, Canvasback, Redhead, Lesser Scaup, Surf Scoter, Long-tailed Duck, Bufflehead, Common Goldeneye, Hooded Merganser, Common Merganser, Red-breasted Merganser, Ruddy Duck, Common Loon, Pied-billed Grebe, Horned Grebe	Max. numbers over 100 ind. of each species seen during winter	Winter	Nomination, Bob Sundell
Congregations-Individual Species	Pied-billed Grebe	Max. daily fall migration: 20 ind. in 1998; 22 in 1999, 48 in 2000, 120 in 2001, 67 in 2002, 5 in 2003	Fall	Dick Miga pers. comm. 2004

Description: Located in the southwest corner of the state, Chautauqua Lake is approximately 16 miles long, with a maximum width of two miles. The village of Mayville sits on the northern end and Jamestown lies on the southern end, with I-86 intersecting the lake on a high elevation bridge. The lake is owned by the State of New York and lakeshore ownership includes private, municipal, and state (administered by NYS OPRHP and NYS DEC).

Birds: This site is an important stopover location for migrant birds, particularly waterfowl. At least 270 species have been documented. Maximum numbers of selected species that have been documented over the last 20 years include 3,000 Tundra Swans, 1,000 plus Canvasbacks, 800 Redheads, 400 Lesser Scaup, 100 plus Surf Scoters, 120 Long-tailed Ducks, 650 Buffleheads, 800 Common Goldeneyes, 1,200 Hooded Mergansers, 600 Common Mergansers, 400 Red-breasted Mergansers, 1,200 Ruddy Ducks, 40 Common Loons, 120 Pied-billed Grebes, 250 Horned Grebes, 4,500 American Coots, 50 Semipalmated Plovers, 200 Killdeer, 110 Lesser Yellowlegs, 80 Semipalmated Sandpipers, 250 Bonaparte's Gulls, 8,000 Ring-billed Gulls, 3,000 Herring Gulls, 87 Common Terns, and 23 Black Terns. This site has regularly supported over 1% of the estimated state wintering population of Pied-billed Grebes.

Conservation: This site is listed in the 2002 State Open Space Conservation Plan as a priority site under the project name Chautauqua Lake Access, Shore Lands, and Vistas. Although recreational boaters and fishermen heavily use Chautauqua Lake, no disturbance problems have been reported. Pollution, including non-point source agricultural runoff and recreational boating, could negatively impact this aquatic ecosystem. Very little land around the lake is in public ownership and only 11% of the lakeshore habitat remains undeveloped. There are 461 acres of wetland habitat at the outlet; 210 are owned by the city of Jamestown and 251 acres, a mercantile zone within the town of Ellicott, are privately owned. The Chautauqua Watershed Conservancy (CWC) and the NYS DEC have successfully protected some of the shoreline habitat. CWC received a North American Wetland Conservation Act grant in 2004 to acquire and restore 50 acres at southern end. Continued monitoring of waterfowl on the lake is needed.

Connecticut Hill Area

Multiple municipalities, Schuyler and Tompkins Counties

26,500 acres
1,085-1,975' elevation

42.3402°N
76.6917°W

IBA Criteria Met

Criterion	Species	Data	Season	Source
Species at Risk	Northern Goshawk	Extent of habitat and existing data strongly suggest that the threshold is being met [1]; 2 confirmed pairs, possibly up to 4 pairs [2]	Breeding	1. Technical committee consensus; 2. Stefan Hames pers. comm. 2004
Responsibility Species Assemblage-Forest	Sharp-shinned Hawk, Black-billed Cuckoo, Northern Flicker, Eastern Wood-Pewee, Least Flycatcher, Blue-gray Gnatcatcher, Wood Thrush, Black-throated Blue Warbler, Black-and-white Warbler, Louisiana Waterthrush, Hooded Warbler, Canada Warbler, Scarlet Tanager, Rose-breasted Grosbeak	Breeds	Breeding	NY BBA 2000

Description:

This site is one of the largest high elevation forested areas in the region. Much of area was marginal farmland purchased by the state in the 1930's and 1940's and allowed to reforest. A historic 10-year study of the population ecology of Ruffed Grouse occurred here. The area is traversed by a number of dirt roads and by the Finger Lakes hiking trail. The core property is administered by the NYS DEC (over 11,000 acres) and the surrounding lands are privately owned. According to the NY GAP land cover data, almost 95% of site is forest habitat, which includes Appalachian oak pine, sugar maple mesic, oak, successional hardwood, evergreen northern hardwood, evergreen plantation, and deciduous wetland forests.

Birds: This extensive forest habitat supports characteristic species, including the Sharp-shinned Hawk, Black-billed Cuckoo, Northern Flicker, Eastern Wood-Pewee, Least Flycatcher, Blue-gray Gnatcatcher, Wood Thrush, Black-throated Blue Warbler, Black-and-white Warbler, Louisiana Waterthrush, Hooded Warbler, Canada Warbler, Scarlet Tanager, and Rose-breasted Grosbeak. At-risk species breeding include the Cooper's Hawk (confirmed), Northern Goshawk (confirmed), Red-shouldered Hawk (confirmed), American Woodcock (probable), Willow Flycatcher (probable), Wood Thrush (probable), Blue-winged Warbler (possible), and Canada Warbler (probable).

Conservation: Regular bird monitoring should be encouraged. Permanent protection of the remaining privately owned portions is needed to prevent their development and conversion to non-forest uses. Options include public or land trust acquisition, purchase of conservation easements, and sustainable forestry agreements.

Cowaselon Creek Watershed Area

Multiple municipalities, Onondaga, Oneida, and Madison Counties

46,500 acres
360-490' elevation

43.1255°N
75.9024°W

IBA Criteria Met

Criterion	*Species*	*Data*	*Season*	*Source*
Species at Risk	Cerulean Warbler	5-10 pairs in 2001 and 2002	Breeding	Matt Young pers. comm. 2004
Responsibility Species Assemblage-Forest	Sharp-shinned Hawk, Black-billed Cuckoo, Eastern Wood-Pewee, Wood Thrush, Rose-breasted Grosbeak, Baltimore Oriole	Breeds	Breeding	NY BBA 2000
Responsibility Species Assemblage-Grassland	Killdeer, Upland Sandpiper, Bobolink, Eastern Meadowlark	Breeds	Breeding	NY BBA 2000 and Matt Young pers. comm. 2004

Description: This site is found in the Great Lakes Plains eco-zone just south of Oneida Lake. The area includes abandoned agricultural lands, forests, and recently constructed wetlands. According to the NY GAP land cover data, approximately 30% of this site is forest habitat, which includes sugar maple mesic, oak, successional hardwood, evergreen northern hardwood, and deciduous wetland forests; 20% of this site is open habitat, which includes cropland and old/field pastures. The site is owned by the Great Swamp Conservancy, Inc. (30 plus acres), United States Fish and Wildlife Service (USFWS), NYS DEC (the 3,787-acre Cicero Swamp Wildlife Management Area), Save the County Land Trust (73 acres), and private landowners. Many of the wetlands were built by the Natural Resources Conservation Service (NRCS) as part of the Wetlands Reserve Program, and NRCS holds easements on a number of parcels included in this site.

Birds: The forested habitat at this site is relatively intact compared to other forests in the region. It supports characteristic breeding species, including the Sharp-shinned Hawk, Black-billed Cuckoo, Eastern Wood-Pewee, Wood Thrush, Rose-breasted Grosbeak, and Baltimore

Oriole. The grassland habitat at this site is also relatively intact compared to other grasslands in the region. It supports characteristic breeding species, including the Killdeer, Upland Sandpiper, Bobolink, and Eastern Meadowlark. Habitat restoration efforts have increased the diversity and density of breeding birds. Rare to uncommon wetland breeders include the Blue-winged Teal, Northern Shoveler (nested in 2002), Green-winged Teal, Common Moorhen, American Coot, Sora, Black-billed Cuckoo, and Yellow-billed Cuckoo. Historically, Henslow's Sparrows bred in the area (Smith Road) and some may still be supported at the site. A large Long-eared Owl roost (more than 10 birds) has been noted in the area. In spring and fall, waterfowl numbers average between 2,000-4,000, with numbers peaking in mornings and evenings. The site also supports a number of at-risk species, including the American Black Duck (one pair), Pied-billed Grebe (at least two pairs), American Bittern (at least one pair), Least Bittern (at least one pair), Osprey (confirmed breeder), Bald Eagle (two individuals during migration), Northern Harrier (at least two pairs), Sharp-shinned Hawk (confirmed breeder), Cooper's Hawk (two pairs in 2001-2002), Red-shouldered Hawk (probable breeder), Virginia Rail (six to eight breeding), Upland Sandpiper (two pairs in 2000), American Woodcock (10 pairs), Willow Flycatcher (confirmed breeder), Horned Lark (two pairs), Sedge Wren (probable breeder), Wood Thrush (10-20 pairs), Blue-winged Warbler (two to eight pairs), Cerulean Warbler (five-10 pairs), and Vesper Sparrow (two pairs). High numbers of shorebirds also use the area during fall migration. Documented numbers include 50 plus American Golden Plovers, 50-80 White-rumped Sandpipers, and three Buff-breasted Sandpipers.

Conservation: The Great Swamp Conservancy, USFWS, and NRCS have undertaken conservation measures at this site. Wetland easements exist at more than 50 sites throughout the area and additional wetlands are being constructed. Hunting is allowed at this site. The Great Swamp Conservancy owns a nature and education center at the site and is involved with a number of conservation activities, including working with local landowners to protect and manage their lands. They are also monitoring the breeding success of Mallards, Tree Swallows, and Eastern Bluebirds, and re-establishing Ring-necked Pheasants and Northern Bobwhite. Monitoring of at-risk species is encouraged.

Derby Hill Bird Observatory

Mexico, Oswego County

70 acres
265-295' elevation

43.5261°N
76.2406°W

IBA Criteria Met

Criterion	Species	Data	Season	Source
Congregations-Raptors	Mixed species	Average 32,149, maximum 41,315 ind. observed during the 2002-2004 seasons	Spring	Derby Hill Bird Observatory

Broad-winged Hawk

Description: This site is located on the southeast corner of Lake Ontario; the observatory is one of the highest points in the area. In spring, as birds migrate north along the southern shore of Lake Ontario, they follow the shoreline instead of flying over the lake, which funnels the birds over Derby Hill. A broad open field atop the hill provides a wide expanse for viewing migrating hawks. The site also has mixed deciduous woods and a small marsh. Onondaga Audubon Society owns the sanctuary; the surrounding land is privately owned.

Birds: Derby Hill is a well-known spring hawk concentration site and has been monitored annually since 1963. The average number of hawks counted each spring from 1979-1996 was 43,293, with a maximum of 66,139. At least 20 diurnal raptor species have been recorded here, including annual spring averages of 2,997 Turkey Vultures (maximum 7,537), 406 Osprey (maximum 692), 37 Bald Eagles (maximum 101), 780 Northern Harriers (maximum 1,554), 5,936 Sharp-shinned Hawks (maximum 11,582), 543 Cooper's Hawks (maximum 1,176), 70 Northern Goshawks (maximum 174), 950 Red-shouldered Hawks (maximum 1,805), 22,449 Broad-winged Hawks (maximum 40,108), 7,979 Red-tailed Hawks (maximum 19,531), 396 Rough-legged Hawks (maximum 656), 24 Golden Eagles (maximum 55), 497 American Kestrels (maximum 931), 19 Merlins (maximum 53), and four Peregrine Falcons (maximum 12). The site is also an important spring stopover site and concentration point for migrating passerines. Offshore waterfowl, particularly sea ducks, diving ducks, and gulls, regularly number in the thousands. The site is one of few viewing locations for fall jaeger flights, with more than 200 (mostly Parasitic) counted on one day in October 1979. The Sage Creek Marsh has supported breeding American Bitterns and Black Terns (twice in the last five years—one pair in 1999 and two pairs in 2002), and Least Bitterns have been heard.

Conservation: Sage Creek Marsh is threatened by the invasion of purple loosestrife. Erosion of the bluff, particularly during winter storms and spring thaw, is causing the loss of overlook property. During the first round of IBA site identifications, this site was recognized under the research criterion because a long-term monitoring project is based there.

Dunkirk Harbor/Point Gratiot

Dunkirk, Chautauqua County

755 acres
565-590' elevation

42.4931°N
79.3392°W

IBA Criteria Met

Criterion	Species	Data	Season	Source
Species at Risk	Common Tern	Maximum one-day counts: 63 ind. in 1998, 10 in 1999, 35 in 2000, 104 in 2001, 2 in 2001, 17 in 2002, 13 in 2002, 8 in 2002, 45 in 2003	Migration	Dick Miga pers. comm. 2004
Congregations-Waterfowl	Mixed species flocks	One-day count totals: 2,162 ind. in 1988, 3,109 in 1987, 2,297 in 1986	Winter	Harbor Watch at Dunkirk Harbor
Congregations-Gulls	Laughing Gull, Bonaparte's Gull, Ring-billed Gull, Herring Gull, Glaucous Gull, Great Black-backed Gull	One-day count totals: 9,014 ind. in 1988, 15,058 in 1987, 15,029 in 1986	Winter	Harbor Watch at Dunkirk Harbor
Congregations-Individual Species	Red-breasted Merganser	Site has supported over 1% of estimated winter population; 2,000 ind. in 1988, 3,000 in 1987, 1,000 in 1986	Winter	Harbor Watch at Dunkirk Harbor

Description: Located in the southwestern corner of New York State on Lake Erie's southeastern shoreline at the city of Dunkirk, Dunkirk Harbor and Point Gratiot are approximately 53 miles southwest of Buffalo. The harbor is formed and protected by Point Gratiot peninsula and is a popular boat launching and fishing site from spring through fall. The NRG Energy Inc. company operates a power plant that discharges warm water into the harbor, keeping it ice-free during winter. This proves attractive for gulls, ducks, and other waterbirds. Point Gratiot has a shoreline with beaches and bluffs, and also houses the power plant. Much of this site is municipally owned, but significant acreages are corporately owned and some land is private.

Birds: Dunkirk Harbor supports a significant abundance and diversity of waterbirds from fall through spring. Maximums include 300 Redheads (1986), 300 Greater Scaup (1986), 2,000 Common Mergansers (1986), 5,000 Red-breasted Mergansers (1985), 8,000 Bonaparte's Gulls (1987), and 15,000 Ring-billed Gulls (1986). Because Lake Erie completely freezes over in most winters, the open waters of the harbor attract many wintering waterfowl. The harbor has also hosted many rare and unusual species. Point Gratiot is a well-known migratory stopover site for a great diversity of landbird species. It is one of few locations in the state with breeding Red-headed Woodpeckers. Other at-risk species using the site include migrating Common Loons (daily maximum of nine individuals in 2003) and Pied-billed Grebes (daily maximum of 15 individuals in 1999).

Conservation: Development of boat docks and infrastructure to support more recreational boating activities could impact waterbirds that use the harbor. Water quality in the harbor is regularly monitored. Waterbird monitoring has been carried out by the Lake Erie Bird Club and should continue. Point Gratiot is managed as a public use/recreation area. Natural shrub undergrowth is periodically removed from portions of the site to the detriment of migrant passerines. Better inventory and monitoring of migrant bird use of Point Gratiot is needed.

Finger Lakes National Forest

Multiple municipalities, Schuyler and Seneca Counties

16,000 acres
980-1,780' elevation

42.5166°N
76.7920°W

IBA Criteria Met

Criterion	*Species*	*Data*	*Season*	*Source*
Species at Risk	American Woodcock	Confirmed breeder, found in 12% of administrative compartments	Breeding	David Clayton Grove pers. comm. 2003
Species at Risk	Wood Thrush	Confirmed breeder found in approximately 55% of site	Breeding	David Clayton Grove pers. comm. 2003
Species at Risk	Blue-winged Warbler	Confirmed breeder, found in approximately 38% of site	Breeding	David Clayton Grove pers. comm. 2003
Species at Risk	Henslow's Sparrow	Breeds most years, declining in recent years [1]; 5-6 pairs in 1997 [2]	Breeding	1. David Clayton Grove pers. comm. 2003; 2. IBA Nomination, David Clayton Grove
Responsibility Species Assemblage-Grassland	Killdeer, Bobolink, Eastern Meadowlark	Breeds	Breeding	NY BBA 2000

Description:

This site consists of a large area of abandoned farmland located on a ridge between Seneca and Cayuga Lakes in the Finger Lakes Region. Between 1938 and 1941 over 100 farms were acquired, and the U.S. Forest Service now owns the site. The area is made up of three successional communities: nearly 6,000 acres of grassland, about 2,500 acres of shrubland, and over 7,500 acres of deciduous upland forest. The grasslands are kept open through a cattle grazing lease program. According to the NY GAP land cover data, approximately 20% of this site is open habitat, which includes cropland and old field/pastureland. The site is now managed as a multiple use area for recreation, grazing, wildlife conservation, timber, education, and research.

Birds: The site is particularly important as a grassland bird breeding site for Northern Harriers (annual breeders), Upland Sandpipers (though none since 1990), Horned Larks, Sedge Wrens (one pair in 1997), Vesper Sparrows (one pair in 1997, presumed breeding), Savannah Sparrows, Grasshopper Sparrows (confirmed nesting), Henslow's Sparrows (30-45 pairs in 1997, not confirmed recently), Bobolinks, and Eastern Meadowlarks. Additional at-risk species found at the site include the Sharp-shinned Hawk (several nesting records), Cooper's Hawk (several nesting records), Northern Goshawk (at least one pair), Red-shouldered Hawk (possible breeder), Short-eared Owl (migrant), Willow Flycatcher (known nester), and Wood Thrush (confirmed nester). Clay-colored sparrows were seen at the site in 2004. The site hosts a great variety of breeding birds (119 species) among its diverse habitats, including characteristic deciduous forest breeding birds. The site is also becoming important to shrub-breeding species such as the Blue-winged Warbler, Eastern Towhee, and Field Sparrow.

Conservation: The grassland and shrub habitats at this site would eventually grow up into forest without active management. Currently, grassland areas are kept open through a combination of grazing, mowing, hand cutting, and in some areas, burning. Groups including hikers, horseback riders, mountain bikers, hunters, and campers use the area. The levels of such use should continue to be monitored to prevent negative impacts on birds like the Northern Goshawk. The Finger Lakes Land Trust is working to secure conservation easements on properties adjoining the National Forest. The possibility of oil and gas exploration was considered in 2002. In August 2002, the Senate Energy and Natural Resources Committee passed a bill to prohibit oil and gas drilling in the Finger Lakes National Forest in the State of New York. The bill is now before the full Senate and would permanently ban permitting or leasing for oil or gas drilling in the Finger Lakes National Forest. Bird research, including inventory and monitoring, should continue at the site. During the first round of IBA site identifications, this site was recognized under the research criterion because a long-term monitoring project is based there.

Galen Marsh

Galen, Wayne County

730 acres
395' elevation

43.0684°N
76.9108°W

IBA Criteria Met

Criterion	*Species*	*Data*	*Season*	*Source*
Species at Risk	Cerulean Warbler	10-25 pairs breed	Breeding	Nomination, Dominic Sherony

Description: This site includes upland, marsh, and wetland habitats within the Galen Marsh Wildlife Management Area and surrounding private lands. River Road runs north to south along the Clyde River, which flows through the site. Current management provides protection to a portion of the Galen Marsh, locally known as the Marengo Swamp. Farming and the harvesting of forest products have been the predominant land use since the settlement of the area and these activities continue today.

Birds: This site is an important breeding area for the at-risk Cerulean Warbler. Other breeding warblers in the region include Blue Winged, Black-throated Green, Black-and-white, Mourning and Hooded Warblers, American Redstarts, Northern Waterthrushes, and Common Yellowthroats.

Conservation: Forest management should be conducted in a way that is compatible with the habitat requirements of Cerulean Warblers. Protection of existing forests from fragmentation, and restoration of additional forests are also desirable.

Greater Summerhill Area

Locke and Moravia, Cayuga County

29,000 acres
725-1,750' elevation

42.6716°N
76.3616°W

IBA Criteria Met

Criterion	*Species*	*Data*	*Season*	*Source*
Species at Risk	American Woodcock	35+ ind.	Breeding	Nomination, Matt Young
Species at Risk	Wood Thrush	25+ ind.	Breeding	Nomination, Matt Young
Species at Risk	Blue-winged Warbler	10+ singing males	Breeding	Nomination, Matt Young
Species at Risk	Canada Warbler	30+ ind.	Breeding	Nomination, Matt Young

Description: This site stretches from Summerhill State Forest north to the wetlands on the southern end of Owasco Lake. Habitats include forest, shrub, agricultural grassland, and wetland areas. The site includes state (Summerhill State Forest and Fillmore Glen State Park), non-profit (The Finger Lakes Land Trust's Dorothy McIlroy Bird Sanctuary), and privately owned lands.

Birds: This site supports 125 probable nesting species, including 22 species of nesting warblers. At-risk species supported at the site include the American Black Duck (one pair), American Bittern (at least one), Least Bittern (one calling male one year), Northern Harrier (at least one pair), Sharp-shinned Hawk (at least two pairs), Cooper's Hawk (at least two pairs), Northern Goshawk (one pair occasionally), Red-shouldered Hawk (one pair occasionally), American Woodcock (35 plus), Willow Flycatcher (10 plus), Horned Lark (two plus), Wood Thrush (25 plus), Blue-winged Warbler (10 plus singing males), Prairie Warbler (one), Cerulean Warbler (five plus), Canada Warbler (30 plus, very high densities), Yellow-breasted Chat (one), Vesper Sparrow (one), and Grasshopper Sparrow (one). Additional noteworthy species include the Virginia Rail, Sora, Common Moorhen, American Coot, Acadian Flycatcher (six to eight), Common Raven (at least one pair), Red Crossbill, White-winged Crossbill, and Pine Siskin.

Conservation: The Finger Lakes Land Trust is active in land protection within this site. NYS OPRHP and NYS DEC should assess habitat conditions and identify options for protection and management.

Hamlin Beach State Park

Hamlin, Monroe County

1,000 acres
240-295' elevation

43.3607°N
77.9547°W

IBA Criteria Met

Criterion	*Species*	*Data*	*Season*	*Source*
Congregations-Waterfowl	Mixed flocks	Hundreds of waterfowl can be observed sitting on the surface, feeding, resting, and using the area as a stopover. Many thousands are seen migrating annually.	Migration	Nomination, Bob Marcotte

Bonaparte's Gull

Description: Hamlin Beach State Park, administered by NYS OPRHP, occupies slopes on the south shore of Lake Ontario, west of Rochester. The site includes a marsh at the east end, extensive pine woods in a camping area in the middle of the park, and beaches protected by a series of stone jetties along the shore. The park occupies a portion of shoreline that extends farther north into the lake than the Rochester shore, making it a good site for waterfowl viewing. An organized daily fall lake watch was sponsored by the Braddock Bay Raptor Research group and operated from 1993 to 1999. Since that time, the lake watch has been carried out several days per week throughout the fall, winter, and spring. The New York State Ornithological Association published the results of the 1993-99 seasons in 2001. The site is a premier location for monitoring waterfowl and gulls migrating along the lake; thousands have been recorded flying over during migration. Annual fall migration observers record about 300,000 birds.

Birds: This is a premier spot for migrating waterfowl. In fall 1997, observers tallied 44,167 Greater Scaup (one-day maximum of 2,850), 76,704 White-winged Scoters (one-day maximum of 8,631), 12,980 Black Scoters (one-day maximum of 11,635), 20,420 Long-tailed Ducks (one-day maximum of 4,582), 10,877 Common Goldeneyes (one-day maximum of 1,156), 31,111 Red-breasted Mergansers (one-day maximum of 7,885), 19,841 Red-throated Loons (one-day maximum of 2,777), 16,846 Common Loons (one-day maximum of 2,982), and 32,081 Bonaparte's Gulls (one-day maximum of 5,750) flying over the site. Black Terns have bred here (three pairs in 1989). The site often attracts winter finches, including Red and White-winged Crossbills, Common Redpolls, Pine Siskins, and Pine Grosbeaks.

Conservation: Monitoring of waterfowl numbers should continue.

Happy Valley Wildlife Management Area

Multiple municipalities, Oswego County

9,600 acres
580-690' elevation

43.4471°N
75.9969°W

IBA Criteria Met

Criterion	*Species*	*Data*	*Season*	*Source*
Species at Risk	Red-shouldered Hawk	Nesting attempts within WMA and surrounding area: 9 in 1985, 15 in 1986, and 15 in 1987 [1]; 5 pairs within the WMA in 1994 [2]	Breeding	1. Johnson and Chambers 1994; 2. Scott Crocoll NYS DEC pers. comm. 2004

Red-shouldered Hawk

Description: This site includes the Happy Valley Wildlife Management Area (WMA) and surrounding forests. It supports a variety of ecological communities, plants, and wildlife. The area is relatively flat and is drained by two watersheds, the Little Salmon River and the Oswego River, both of which eventually drain into Lake Ontario. The site is characterized by second growth northern hardwood forests, dominated by red maple (*Acer rubrum*) and mixed hardwood-conifer stands. Two significant ecological communities, spotted turtle (*Clemmys guttata*), and four rare plants were documented during a 1994 biodiversity inventory. The fens at St. Mary's Pond and Long Pond are among the finest examples of this rare community in the state.

Birds: This area supports characteristic breeding bird communities of forested wetlands and mixed deciduous/coniferous forests. Breeding species include Pied-billed Grebes (confirmed breeder), Sharp-shinned Hawk (confirmed breeder), Northern Goshawk (confirmed pair), Broad-winged Hawk (confirmed pair), Ruffed Grouse, American Woodcock, Black-billed Cuckoo, Yellow-bellied Sapsucker, Barred Owl (confirmed breeder), Eastern Wood-Pewee, Least Flycatcher, Great Crested Flycatcher, Blue-headed Vireo, Veery, Bicknell's Thrush, Wood Thrush, Chestnut-sided Warbler, Black-throated Blue Warbler, Black-throated Green Warbler, American Redstart, Ovenbird, Canada Warbler, Scarlet Tanager, Rose-breasted Grosbeak, and Purple Finch. Red-shouldered Hawks breed here in relatively high densities (average of 11 nesting attempts each year).

Conservation: This area is currently managed primarily for wildlife conservation and wildlife related recreational activities (hunting and birding). Logging also takes place at the site; logging plans should take Red-shouldered Hawk nest locations into consideration. Inventory and monitoring of Red-shouldered Hawks and other at-risk species are needed.

Hemlock and Canadice Lakes

Multiple municipalities, Livingston and Ontario Counties

28,000 acres
885-2,070' elevation

42.7094°N
77.5905°W

IBA Criteria Met

Criterion	*Species*	*Data*	*Season*	*Source*
Species at Risk	Bald Eagle	2 pairs in 2000-2003	Breeding	NY Natural Heritage Biodiversity Databases
Responsibility Species Assemblage-Forest	Sharp-shinned Hawk, Eastern Wood-Pewee, Rose-breasted Grosbeak, Wood Thrush, Baltimore Oriole	Breeds	Breeding	NY BBA 2000

Description: Part of the glacially created Finger Lakes, Hemlock (7.5 miles long) and Canadice (3.2 miles long) are long, thin lakes that run north to south. Both lakes have steep western shores and are ringed with extensive forested buffers, beyond which lies open/cleared lands. According to the NY GAP land cover data, 65% of this site is forest habitat, which includes deciduous wetland, evergreen northern hardwood, evergreen plantation, successional hardwood, and sugar maple mesic forests. These lakes provide drinking water to the City of Rochester and several other towns.

Birds: In the 1970s, Hemlock Lake was the site of the last "wild" Bald Eagle nest in the state. Descendants of that pair continue to nest here. The site hosts a great diversity of species characteristically found in old field, shrub, and forest habitats. (126 species have been tallied as potential or confirmed breeders according to the NY Breeding Bird Atlas.) The site also supports breeding populations of a number of other at-risk species, including Northern Harriers, Sharp-shinned Hawks, Cooper's Hawks, Northern Goshawks, Red-shouldered Hawks, Red-headed Woodpeckers, Willow Flycatchers, Horned Larks, Blue-winged Warblers, Golden-winged Warblers, Cerulean Warblers, Vesper Sparrows, Grasshopper Sparrows, and Henslow's Sparrows (present in 1997).

View looking north on Hemlock Lake

Conservation: This site is listed in the 2002 State Open Space Conservation Plan as a priority site under the project name Western Finger Lakes: Conesus, Hemlock, Canadice and Honeoye. Currently the city of Rochester owns 7,100 acres of land that buffer both Hemlock and Canadice Lakes—the only Finger Lakes with no shoreline development. Rochester protects the public water supply and allows community recreation via free permits, with restrictions on the type of use, both on land and water. The primary threat outside of the city-owned shoreline is the potential for increased residential development. The Finger Lakes Land Trust and The Nature Conservancy have been working to protect land in this area. The Nature Conservancy owns about 400 acres and the Finger Lakes Land Trust holds three easements, which complement the city-owned parcels. The NYS DEC has a long-standing cooperative program with the city of Rochester to monitor the last "wild" Bald Eagle nest in the state. During this time, young were imported to enhance the nest's production when the old pair had difficulty breeding. A Christmas Bird Count has been carried out annually within the area for 45 years. In addition, a breeding bird census was done in the same area for five years, and regular Breeding Bird Survey routes run through the area. Continued inventory and monitoring of at-risk species are needed. There are concerns about hemlock wooly adelgid (*Adelges tsugae*) spreading in this area. Forest management should be conducted in a way that is compatible with the habitat requirements of forest bird species. During the first round of IBA site identifications, this site was recognized under the research criterion because a long-term monitoring project is based there.

High Tor Wildlife Management Area

Multiple municipalities, Ontario and Yates Counties

6,100 acres
680-1,835' elevation

42.6352°N
77.3522°W

IBA Criteria Met

Criterion	*Species*	*Data*	*Season*	*Source*
Species at Risk	Pied-billed Grebe	Breeds	Breeding	Nomination, June Summers
Species at Risk	American Bittern	Breeds	Breeding	Nomination, June Summers
Species at Risk	Least Bittern	Breeds	Breeding	Nomination, June Summers
Species at Risk	Cooper's Hawk	Breeds	Breeding	Nomination, June Summers
Species at Risk	Northern Goshawk	Breeds	Breeding	Nomination, June Summers

Description: This site includes three diverse habitats administered by NYS DEC. The largest section, approximately 3,400 acres, is a steep, wooded terrain with a number of human-made impoundments, several of which are stocked with trout. The second is a 1,700-acre freshwater marsh that borders the southern end of Canandaigua Lake and is known for its rainbow trout (*Oncorhynchus mykiss*) spawning runs. The third, a 1,000-acre area, is located east of the southern end of Canandaigua Lake and includes old fields and steep, wooded hillsides.

Birds: The 1,700-acre marsh and wooded swamp area at the south end of Canandaigua Lake supports common cattail (*Typha latifolia*) and many characteristic wetland birds. Documented species include breeding Pied-billed Grebes, American Bitterns, Least Bitterns, Virginia Rails, Soras, and Common Moorhens. One Prothonotary Warbler was noted in the summer of 1997. The remaining 4,400 acres of upland habitat support breeding Cooper's Hawks, Northern Goshawks, Black-billed Cuckoos, Yellow-bellied Sapsuckers, Yellow-throated Vireos, Blue-headed Vireos, Red-eyed Vireos, Common Ravens, Red-breasted Nuthatches, Carolina Wrens, Hermit Thrushes, Wood Thrushes, Brown Thrashers, Blue-winged Warblers, Chestnut-sided Warblers, Magnolia Warblers, Black-throated Blue Warblers, Yellow-rumped Warblers, Black-throated Green Warblers, Blackburnian Warblers, Ovenbirds, Mourning Warblers, Hooded Warblers, Canada Warblers, Scarlet Tanagers, Field Sparrows, Dark-eyed Juncos, Indigo Buntings, Bobolinks, Eastern Meadowlarks, Orchard Orioles, and Purple Finches (all noted in 1997 breeding season).

Conservation: This site is listed in the 2002 State Open Space Conservation Plan as a priority site under the project name High Tor/Bristol Hills. This site has been designated as a state Bird Conservation Area. The Finger Lakes Land Trust owns 230 acres adjoining the northern High Tor area, and there are ongoing efforts to acquire additional acreage. The site is managed as a wildlife conservation area, but development of hunting camps, summer homes, and year-round homes takes place along the borders of the area. A management plan for the site has been developed; however, it is focused on game species management, and little mention is made of non-game species. Monitoring of at-risk species is needed to better understand bird use and to determine whether this site continues to meet IBA criteria.

Iroquois NWR/Oak Orchard and Tonawanda WMAs

Multiple municipalities, Genesee and Orleans Counties

20,000 acres
595-650' elevation

43.1240°N
78.3941°W

IBA Criteria Met

Criterion	*Species*	*Data*	*Season*	*Source*
Species at Risk	Pied-billed Grebe	27 ind. in 1995, 27 in 1996, 10 in 1997, 6 in 1998, 6 in 1999, 6 in 2000	Breeding	BSC Marshbird Monitoring Program
Species at Risk	Least Bittern	5 ind. in 2001, 3 in 2000, 4 in 1999, 1997, and 1996	Breeding	BSC Marshbird Monitoring Program
Species at Risk	Bald Eagle	3 pairs in 2001-2002	Breeding	NY Natural Heritage Biodiversity Databases
Species at Risk	American Woodcock	Average of 53 singing males/year at refuge	Breeding	Paul Hess pers. comm. 2004
Species at Risk	Black Tern	27 ind. in 1995, 20 in 1996, 8 in 1997, 4 in 1998, 1 in 1999, 13 in 2000, 12 in 2001	Breeding	BSC Marshbird Monitoring Program
Species at Risk	Sedge Wren	4-5 singing males/year, breeds	Breeding	Paul Hess pers. comm. 2004
Species at Risk	Cerulean Warbler	16 ind. in 1997, 94 in 1998, 4 in 1999	Breeding	Cerulean Warbler Atlas Project
Responsibility Species Assemblage-Grassland	Killdeer, Bobolink, Eastern Meadowlark	Breed	Breeding	NY BBA 2000
Congregations-Waterfowl	Mixed species	49,348 ind. in 2004, 32,002 in 2003, 41,954 in 2002, 51,122 in 2001, 62,776 in 2000		Paul Hess pers. comm. 2004
Congregations-Wading Birds	Great Blue Heron	270 nests in 1995, 44 in 1996, 86 in 1998, 292 in 1999, 321 in 2000, 269 in 2001, 211 in 2002, 399 in 2003, 324 in 2004		Paul Hess pers. comm. 2004

Description: Otherwise known as the "Alabama Swamp," this complex encompasses nearly 20,000 acres of protected wildlife habitat; approximately 70% of the site is wetland habitat. There are also grassland and forest habitats. Approximately 1,400 acres are managed as grasslands. The surrounding area is mainly agricultural. Iroquois National Wildlife Refuge (10,800 acres) is owned and managed by the U.S. Fish and Wildlife Service, while Oak Orchard (2,500 acres) and Tonawanda (5,600 acres) Wildlife Management Areas are administered by the NYS DEC.

Birds: This important site for breeding and migratory waterfowl hosts an estimated 100,000 waterfowl every spring. The area also supports many at-risk species, including the Brant (casual migrant), American Black Duck (breeds), Pied-billed Grebe (breeds), American Bittern (breeds), Least Bittern (breeds), Osprey (breeds), Bald Eagle (breeds), Northern Harrier (breeds), Sharp-shinned Hawk (breeds), Cooper's Hawk (breeds), Upland Sandpiper (two singing males in 2001), American Woodcock (breeds), Black Tern (breeds), Short-eared Owl (winters), Whip-poor-will (casual migrant), Red-headed Woodpecker (breeds, year round resident), Willow Flycatcher (common breeder), Sedge Wren (breeds), Wood Thrush (common breeder), Blue-winged Warbler (breeds), Golden-winged Warbler (rare breeder), Cerulean Warbler (breeds), Prothonontary Warbler (rare breeder, one individual in 2001), Yellow-breasted Chat (three males during the 2002 breeding season), Grasshopper Sparrow (breeds), Henslow's Sparrow (three singing males in 2001 and 2002), and Rusty Blackbird (migrant). All migratory species that breed at this site also use it during migration. The site has also supported a large Great Blue Heron rookery.

Conservation: Biological controls have successfully reduced the amount of purple loosestrife *(Lythrum salicaria)* on the refuge. Portions of this site have been designated a state Bird Conservation Area. Maintenance of the grassland habitats is encouraged. Continued inventory and monitoring of non-game wetland and grassland birds, especially at-risk species, is recommended. During the first round of IBA site identifications, this site was recognized under the research criterion because it is a site where long-term research and monitoring projects are based.

Keeney Swamp Forest

Birdsall, Allegany County

3,300 acres
1,675-2,025' elevation

42.4147°N
77.9040°W

IBA Criteria Met

Criterion	*Species*	*Data*	*Season*	*Source*
Species at Risk	Northern Harrier	Breeds	Breeding	Nomination, Walt Franklin
Species at Risk	Cooper's Hawk	Breeds	Breeding	Nomination, Walt Franklin
Species at Risk	Northern Goshawk	Breeds	Breeding	Nomination, Walt Franklin
Species at Risk	Red-shouldered Hawk	Breeds	Breeding	Nomination, Walt Franklin
Species at Risk	Red-headed Woodpecker	Breeds	Breeding	Nomination, Walt Franklin
Species at Risk	Wood Thrush	Breeds	Breeding	Nomination, Walt Franklin

Description: This site includes private, county-owned, and state lands (NYS DEC WMA and State Forest). A variety of habitats are found at the site, including extensive wetlands and two forest types: Alleganian and sub-Canadian. The latter lies above 1,800', and contains native stands of balsam fir (*Abies balsamea*). The site supports a diversity of birds and other wildlife.

Birds: The wetlands at this site are known to support migrant Pied-billed Grebes, American Bitterns, Virginia Rails, Soras, Common Moorhens, and Tundra Swans (400 in 1976). Pied-billed Grebes are probable breeders and marsh bird surveys have detected Virginia Rails, Soras, and Common Moorhens. The upland habitats support characteristic breeding species, including the Cooper's Hawk, Northern Goshawk, Red-shouldered Hawk (breeding noted in 1997), Saw-whet Owl, Common Raven, Red-breasted Nuthatch, Golden-crowned Kinglet, Chestnut-sided Warbler, Blackburnian Warbler, Mourning Warbler, Canada Warbler, and White-throated Sparrow. Some grassland habitat remains within the state forest and supports breeding Northern Harriers, Vesper Sparrows, Bobolinks, and Eastern Meadowlarks. Upland Sandpipers have been regularly noted here during spring migration.

Conservation: A 3,000-foot dike that impounds about 105 acres was completed in 2004. This impoundment, along with beaver reoccupation, has increased the amount of marsh habitat significantly. Management plans for the state forest should incorporate the needs of wetland and forest bird species. More inventory and monitoring of at-risk species is needed to better understand bird use and to determine whether this site continues to meet IBA criteria.

Virginia Rail

Letchworth State Park

Multiple municipalities, Livingston and Wyoming Counties

14,000 acres
590-1,380' elevation

42.6531°N
77.9706°W

IBA Criteria Met

Criterion	*Species*	*Data*	*Season*	*Source*
Species at Risk	American Woodcock	10-20 pairs per year	Breeding	Douglas Bassett pers. comm. 2003
Species at Risk	Willow Flycatcher	20-30 pairs per year	Breeding	Douglas Bassett pers. comm. 2003
Species at Risk	Wood Thrush	50-100 pairs per year	Breeding	Douglas Bassett pers. comm. 2003
Species at Risk	Blue-winged Warbler	250-300 pairs	Breeding	Douglas Bassett pers. comm. 2003
Species at Risk	Cerulean Warbler	15-20 pairs per year	Breeding	Douglas Bassett pers. comm. 2003
Species at Risk	Canada Warbler	75-100 pairs per year	Breeding	Douglas Bassett pers. comm. 2003
Species at Risk	Yellow-breasted Chat	Up to 10 pairs documented	Breeding	Douglas Bassett pers. comm. 2003
Species at Risk	Rusty Blackbird	Thousands migrate through	Migration	Douglas Bassett pers. comm. 2003

Description: Called "the Grand Canyon of the East," this area is bisected by the Genesee River. Three dramatic gorges within the park—one 550 feet deep—have been cut by the river, revealing the geologic history of the underlying rock. The park itself is approximately 15 miles long and two miles wide, with three waterfalls of 70-100 feet each, and a canyon six miles long. Mixed forest tops the gorge walls for the length of the park. One fifth of the land is used by the Mt. Morris Flood Control Dam System. The park is administered by NYS OPRHP, but the dam system is controlled by the federal government.

Birds: The forests and grasslands at this site support an exceptional diversity of breeding warblers (25 species), songbirds, and many at-risk species, including breeding American Black Ducks (several pairs every year), Bald Eagles (one pair), Northern Harriers (two pairs in 1993-2003), Cooper's Hawks (three to five pairs in 1993-2003), Northern Goshawks (one to two pairs most years), Red-shouldered Hawks (one to two pairs most years), American Woodcocks (10-20 pairs per year), Short-eared Owls (winters), Red-headed Woodpeckers (one to three pairs most years), Willow Flycatchers (20-30 pairs per year), Wood Thrushes (50-100 pairs per year), Blue-winged Warblers (250-300 pairs), Golden-winged Warblers (one to three pairs most years), Prairie Warblers (fewer than three pairs, in four of the last ten years), Cerulean Warblers (15-20 pairs per year), Canada Warblers (75-100 pairs per year), Grasshopper Sparrows (one to two pairs most years), Henslow's Sparrows (fewer than three pairs, in seven of the last 10 years), and Rusty Blackbirds (thousands migrate through). This is one of the few sites in upstate New York with breeding Yellow-breasted Chats (documented in five of the last 10 years). It also supports a large winter roost of Turkey Vultures.

Conservation: Most of the area is managed in ways that are beneficial to birds. However, changes in water level from the Mt. Morris Federal Flood Control Dam have the potential to negatively impact birds. Succession of grasslands to shrublands should be managed to maintain habitat for Grasshopper Sparrows, Henslow's Sparrows, and other grassland species. The park is a popular recreation site (one million annual visitors). There is a continuing need to assess the impacts of recreation use on bird populations and their habitats. Inventory and monitoring of breeding birds should continue.

Long Pond State Forest

Smithville, Chenango County

3,100 acres
1,110-1,480' elevation

42.4206°N
75.8357°W

IBA Criteria Met

Criterion	Species	Data	Season	Source
Species at Risk	Henslow's Sparrow	6 ind. in 2000; 19 pairs in 1995; 9 ind. in 1993	Breeding	NY Natural Heritage Biodiversity Databases

Description: This site is administered by NYS DEC and consists of a large multiple-use area with a diversity of habitats, including a 117-acre pond accessible to fishermen, 400 acres of grassland and scrubland, and 1,500 acres of mature hardwood-hemlock forest, interspersed with wetlands. Approximately 300 acres of hardwood-hemlock forest have been excluded from timber harvest and contain large hemlocks, some over 125 years old. The grassland portions once belonged to local dairy farms and are now managed by controlled mowing and burning. The area is popular for hunting, fishing, hiking, camping, picnicking, and snowmobiling.

Birds: This site is particularly important as a breeding area for grassland birds, including the Grasshopper Sparrow (three plus pairs in 1996) and Henslow's Sparrow (19 pairs in 1995, seven in 1996). The mature hardwood-hemlock forest supports breeding Red-shouldered Hawks.

Conservation: Portions of this site have been designated as a state Bird Conservation Area. Henslow's Sparrows have been observed within the last ten years, and there is a real need to survey this site further to determine their current numbers. Controlled mowing and burning of the grasslands are part of the site management plan. These activities have been successful in maintaining grassland bird habitat. Continued monitoring of grassland birds is needed to provide feedback for management. The impacts of recreational activities on breeding birds should be considered in future management plans.

Montezuma Wetlands Complex

Multiple municipalities, Cayuga, Seneca, and Wayne Counties

70,000 acres
320-520' elevation

43.0749°N
76.7515°W

IBA Criteria Met

Criterion	*Species*	*Data*	*Season*	*Source*
Species at Risk	Pied-billed Grebe	Confirmed breeding in 3 atlas blocks; probable in a 4th block	Breeding	NY BBA 2000
Species at Risk	Least Bittern	Breeds, 3 ind. at edge of marsh	Breeding	Kevin McGowan pers. comm. 2004
Species at Risk	Osprey	10-12 pairs nesting in complex	Breeding	Tracy Gingrich pers. comm. 2004
Species at Risk	Bald Eagle	3 pairs	Breeding	Tracy Gingrich pers. comm. 2004
Species at Risk	Black Tern	10-20 pairs	Breeding	Tracy Gingrich pers. comm. 2004
Species at Risk	Sedge Wren	5-10 pairs nesting in complex	Breeding	Tracy Gingrich pers. comm. 2004
Species at Risk	Cerulean Warbler	Estimated 60-85 pairs, probably more, in 2004 [1]; 207 ind. in 1997, 104 in 1998, 13 in 1999, 5 in 2002 [2]	Breeding	1. Jim Eckler pers. comm. 2004; 2. Cerulean Warbler Atlas Project
Responsibility Species Assemblage-Forest	Sharp-shinned Hawk, Black-billed Cuckoo, Eastern Wood-Pewee, Wood Thrush, Cerulean Warbler, Rose-breasted Grosbeak, Baltimore Oriole	Breeds	Breeding	NY BBA 2000

Criterion	Species	Data	Season	Source
Congregations-Waterfowl	Mixed species flocks	Over 500,000 Canada Geese pass through the complex during each migration; 15,000 Snow Geese regularly use the area; in late fall American Black Duck numbers peak at over 25,000, and Mallards at 100,000	Migration	Nomination, USFWS
Congregations-Wading Birds	Least Bittern, Great Blue Heron, Great Egret, Green Heron, Black-crowned Night-Heron	Great Blue Herons and Great Egrets can reach 70 ind. at May's Pool [1]; 213 ind. in 2000 [2]	Migration	1. Jillian Liner pers. comm. 2004; 2. Burger 2000
Congregations-Shorebirds	Mixed species flocks	Daily max. of 641 ind. in 2003, 613 in 2002, 726 in 2001	Migration	Audubon NY, Fall surveys at Montezuma

Description: The Montezuma Wetland Complex lies within the heart of the drumlins region of the New York Great Lakes Plain. Broad, flat basins interspersed with classic drumlin glacial formations characterize the area; these oval-shaped hills are generally oriented in a north-south direction with wetland basins in the valleys. The mix of extensive marshes, swamps, upland forests, productive agricultural soils, and varied topography and hydrology create a patchwork of diverse habitats important to many migratory and resident wildlife species. Due to the site's location along the Atlantic Flyway, it provides essential waterfowl feeding and resting areas, and a link between the deep-water habitats of Lake Ontario and the Finger Lakes. According to the NY GAP land cover data, approximately 45% of the site is forested, which includes deciduous wetland, evergreen northern hardwood, evergreen plantation, sugar maple mesic, and successional hardwood forests. Site ownership includes the USFWS (Montezuma National Wildlife Refuge), NYS DEC (6,300-acre Northern Montezuma Wildlife Management Area), and private landowners.

Birds: The wetland habitats found at this site support an abundance and diversity of wetland-dependent species, including one of the largest migratory concentrations of waterfowl in the Northeast. Over 500,000 Canada Geese pass through the complex each migration period. During spring migration, 15,000 Snow Geese regularly use the area. In late fall, American Black Duck numbers peak at more than 25,000, and

Mallards peak at 100,000. Montezuma is one of the most significant stopover and foraging locations for shorebirds in upstate New York, regularly hosting 1,000 or more individuals of 25 species. Many at-risk species breed within the complex, including Pied-billed Grebes, American Bitterns, Least Bitterns, Ospreys, Bald Eagles, Northern Harriers, Cooper's Hawks, Red-shouldered Hawks, Black Terns, Sedge Wrens, and Cerulean Warblers. In addition, the site supports breeding colonies of Great Blue Herons and Black-crowned Night-Herons, and hosts one of the largest fall swallow concentrations in the state, estimated at 50,000-100,000 individuals. In 2003 and 2004, the site also supported the first breeding pair of Sandhill Cranes in the state.

Conservation: A portion of this site is listed in the 2002 State Open Space Conservation Plan as a priority site under the project name Northern Montezuma Wetlands. Portions of this site have been designated as a state Bird Conservation Area. A 24-hour fall birding competition, sponsored by the Friends of the Montezuma Wetlands Complex and Audubon New York and hosted by the USFWS, has raised money for conservation projects for the past eight years. The event supported a MAPS banding station for five years. Currently, a volunteer fall shorebird survey, coordinated by Audubon New York, is conducted at the site annually (2000-2004). Land acquisition by USFWS and NYS DEC continues (750 acres within the complex were acquired by NYS DEC in 1997). Additional land on Route 89 was purchased in 2002 for a grassland restoration project funded by the Friends of the Montezuma Wetlands Complex in 2004. Grassland restoration projects are also underway on both USFWS- and NYS DEC-managed lands. There are some problems involving runoff from nearby croplands into the site's wetlands. The invasion of purple loosestrife (*Lythrum salicaria*) has been a major problem, but an active control program by USFWS and NYS DEC has had some success in decreasing the spread of purple loosestrife and reestablishing common cattail (*Typha latifolia*) in certain areas. Audubon New York is scheduled to open a new Audubon Center at the site in 2006, in partnership with the NYS DEC, the Friends of the Montezuma Wetlands Complex, and the USFWS. This center will work on education, as well as inventory and monitoring projects. During the first round of IBA site identifications, this site was recognized under the research criterion because a long-term monitoring project is based there.

Nation's Road Grasslands

Multiple municipalities, Livingston and Monroe Counties

27,000 acres
490-830' elevation

42.8930°N
77.7916°W

IBA Criteria Met

Criterion	*Species*	*Data*	*Season*	*Source*
Species at Risk	Short-eared Owl	Estimated 12 ind. in 2004, 59 in 1999	Winter	Jim Kimball pers. comm. 2004
Species at Risk	Red-headed Woodpecker	Estimated 6-8 pairs	Breeding	Jim Kimball pers. comm. 2004
Species at Risk	Horned Lark	10 pairs in 1994	Breeding	Jim Kimball pers. comm. 2004
Species at Risk	Cerulean Warbler	Confirmed steady breeding population, 8-12 singing males	Breeding	Jim Kimball pers. comm. 2004
Species at Risk	Grasshopper Sparrow	Estimated 20-30 pairs	Breeding	Jim Kimball pers. comm. 2004
Responsibility Species Assemblage-Grassland	Killdeer, Wilson's Snipe, Henslow's Sparrow, Bobolink, Eastern Meadowlark	Breeds	Breeding	NY BBA 2000

Description: This site includes an exceptional, privately-owned grassland and oak-savanna habitat with a diverse community of breeding and wintering birds. The site lies in the Genesee River Valley among old fields, oak-scattered savanna, and riparian habitat. Active agricultural land lies along the western edge of the site. According to the NY GAP data, approximately 80% of the site is open habitat, which includes cropland and old field/pasture land. The site is home to the second oldest fox hunt in the United States. The Genesee Valley Conservancy holds conservation easements on some of the land within this site.

Birds: Breeding at-risk species include the Northern Harrier (at least two breeding pairs), Sharp-shinned Hawk (at least one pair), Cooper's Hawk (at least one pair), Upland Sandpiper (at least two breeding pairs in 2002-2004), American Woodcock (numerous), Red-headed Woodpecker (six to eight pairs), Willow Flycatcher (numerous), Horned Lark (10 pairs in 1994), Sedge Wren (confirmed nesting pair, plus three singing males in 1998), Wood Thrush (numerous), Blue-winged Warbler (numerous), Yellow-breasted Chat (three singing during the 1998 breeding season; one report since), Vesper Sparrow (five to eight pairs), Savannah Sparrow (40 in 1994), Grasshopper Sparrow (20-30 pairs), Henslow's Sparrow (two to four pairs in 1997, two individuals in 2000), and Bobolink. In winter, the area supports large concentrations of Northern Harriers (four to 10), Rough-legged Hawks (five to 10 in the 1990's), Short-eared Owls (10-59 from 1977-1999), and flocks of Horned Larks that can number in the hundreds. Large numbers of Canada Geese (5,000 plus) feed in the fields during spring migration. The wetlands support healthy numbers of Great Blue Herons, both in migration (maximum 160 in 1995) and breeding (15-20 nests in 2004). A project to follow the reproductive success of Tree Swallows and Eastern Bluebirds, carried out from 1984 to 1989, counted 89 pairs of Tree Swallows and 85 pairs of Eastern Bluebirds in 1989.

Conservation: The primary threat to this site is loss of habitat through the sale and development of land for housing. As the towns of Avon and Geneseo continue to grow, the land has increased in value, making it increasingly tempting for the owners to subdivide and sell. The Genesee Valley Conservancy has protected portions of the site. In-depth inventory of at-risk species is needed, along with a program to educate landowners and the public about the importance of the area to grassland birds.

Niagara River Corridor

Multiple municipalities, Niagara and Erie Counties

98,000 acres
240-680' elevation

43.0654°N
78.9483° W

IBA Criteria Met

Criterion	*Species*	*Data*	*Season*	*Source*
Species at Risk	Common Tern	1,100 nested on the Buffalo Harbor breakwall in 2004	Breeding	Lee Harper pers. comm. 2004
Responsibility Species Assemblage-Shrub	American Woodcock, Willow Flycatcher, Brown Thrasher, Blue-winged Warbler, Eastern Towhee, Field Sparrow	Breed	Breeding	NY BBA 2000
Congregations-Waterfowl	Mixed species	22,431 ind. in 2000, 38,537 in 2001, 11,446 in 2002, 18,938 in 2003, 17,205 in 2004	Winter	NYSOA winter waterfowl counts
Congregations-Waterbirds	Gulls	This site annually supports one of the world's most spectacular gull concentrations, with 19 species recorded and one-day counts of over 100,000 ind.	Winter	Christmas Bird Count, and BOS noteworthy records
Congregations-Waterbirds	Common Tern	1,100 nested on the Buffalo Harbor breakwall in 2004	Breeding	Lee Harper pers. comm. 2004
Congregations-Wading Birds	Great Blue Heron, Great Egret, Black-crowned Night-Heron	233 ind. in 2002, 166 in 2003, 148 in 2004	Breeding	NYS DEC surveys, Mark Kandel
Congregations-Individual Species	Canvasback	Site has supported 31% of state wintering population of Canvasbacks	Winter	NYSOA winter waterfowl counts
Congregations-Individual Species	Common Goldeneye	Site has supported 29% of state wintering population of Common Goldeneyes	Winter	NYSOA winter waterfowl counts

Criterion	*Species*	*Data*	*Season*	*Source*
Congregations-Individual Species	Common Merganser	Site has supported 31% of state wintering population of Common Mergansers	Winter	NYSOA winter waterfowl counts

Description: This site includes the portion of the Niagara River that flows north and northwest for approximately 32 miles from Lake Erie to Lake Ontario, which varies from 110 to 2,200 yards wide. The Upper River flows around Grand Island in eastern and western branches before flowing over Niagara Falls (158-167 feet high) into the Niagara gorge. The gorge is up to 200 feet deep and extends downstream about six miles to the village of Lewiston. The river then flows for another seven miles, between banks ranging from 20-70 feet in height, until it reaches Lake Ontario. There are rapids before and after the falls and a large whirlpool in the lower river. Water depth varies from less than 30 feet in the Upper River, to shallow rapids in the Lower River, to 200 feet in the gorge. The shoreline in some areas of the upper river on the U.S. side is industrially developed and very little natural shoreline remains. In the lower river, the shoreline between the falls and Lewiston is largely undeveloped

Devils Hole on the Niagara River

shrub lands and forests that are protected as state parks. Downstream of Lewiston the shoreline is largely developed, but shrub and forest habitats are still common. According to the NY GAP land cover data, approximately 16% of the site is shrub habitat, which includes old field/ pastures, shrub swamps, successional hardwoods, and successional shrub lands. Portions of the site are administered by NYS OPRHP and NYS DEC, and on Canadian side, by the Niagara Parks Commission, but the bulk of the land is municipal, corporate, or privately owned.

Birds:

The Niagara River annually supports one of the world's most spectacular concentrations of gulls, with 19 species recorded and one-day counts of over 100,000 individuals. The site is particularly noteworthy as a migratory stopover and wintering site for Bonaparte's Gulls, with one-day counts ranging from 10,000-50,000 individuals (2-10% of the world population). One-day Ring-billed Gull counts vary from 10,000-20,000, and one-day Herring Gull counts vary from 10,000-50,000. The river also hosts a remarkable diversity and abundance of waterfowl. Winter NYS DEC aerial surveys show a 22-year average of 2,808 Canvasbacks (31.5% of state wintering population), 2,369 scaup (6% of state wintering population), 2,015 Common Goldeneyes (29% of state wintering population), and 7,527 Common Mergansers (31% of state wintering population). Annual peak numbers range from 2,000-15,000 Canvasbacks, 2,500-15,000 Greater Scaup, 2,300-3,000 Common Goldeneyes, and 2,500-12,000 Common Mergansers. The river also supports breeding colonies of Double-crested Cormorants, Great Blue Herons, Great Egrets, Black-crowned Night-Herons (95-142 pairs), Ring-billed Gulls, Herring Gulls, and Common Terns. The habitats along the river's edge support an exceptional diversity of migratory songbirds during spring and fall migrations. The few remaining marshes, including one at Buckhorn Island State Park, have supported breeding Least Bitterns, Northern Harriers, and Sedge Wrens. Other species at-risk supported at the site include the American Black Duck (breeds), Common Loon (winter), Pied-billed Grebe (confirmed breeder), Cooper's Hawk (confirmed breeder), American Woodcock (probable breeder), Common Nighthawk (probable breeder), Red-headed Woodpecker, Willow Flycatcher (confirmed breeder), Horned Lark (confirmed breeder), Wood Thrush (confirmed breeder), Blue-winged Warbler (probable breeder), and Cerulean Warbler.

Conservation: This site is listed in the 2002 State Open Space Conservation Plan as a priority site under the project name Great Lakes & Niagara River Access, Shore Lands, and Vistas. Portions of this site (Buckhorn Island) are designated as a state Bird Conservation Area. Industrial water pollution on the U.S. side has historically been a major problem along the Niagara River, but cleanup efforts have reduced the levels of most known toxic chemicals. The health of the aquatic ecosystems of the river must continue to be monitored because of the massive numbers of wetland species that rely on those ecosystems. Much of the land along the river has been developed for industrial purposes (especially on the U.S. side of the upper river near Buffalo and Niagara), power generation, and commercial and residential uses. Continued loss of forest and shrub habitats along the river will negatively impact migratory songbirds. The protection of remaining wetland, forest, and shrub habitats along the shoreline should be a priority. Boating activities are a known threat to nesting terns and herons in the upper portions of the river, and there is considerable interest in developing marinas and boat launch facilities along the remaining available shoreline. A comprehensive bird conservation plan for the Niagara River Corridor was developed with input from numerous partners and published in 2002 by the Canadian Nature Federation; efforts should be made to implement the plan's recommendations. Efforts to incorporate bird conservation projects into the Federal Energy Regulatory Commission (FERC) licensing process are encouraged. During the first round of IBA site identifications, this site was recognized under the research criterion because a long-term monitoring project is based there.

Oneida Lake Islands

Multiple municipalities, Oswego and Oneida Counties

2 acres
365' elevation

43.2275°N
75.9970°W

IBA Criteria Met

Criterion	*Species*	*Data*	*Season*	*Source*
Species at Risk	Common Tern	328 pairs in 1996, 351 in 1997, 292 in 1998, 307 in 1999, 311 in 2000, 298 in 2001, 349 in 2002, 449 in 2003	Breeding	New York Cooperative Fish and Wildlife Research Unit, Milo Richmond
Congregations-Waterbird	Common Tern	328 pairs in 1996, 351 in 1997, 292 in 1998, 307 in 1999, 311 in 2000, 298 in 2001, 349 in 2002, 449 in 2003	Breeding	New York Cooperative Fish and Wildlife Research Unit, Milo Richmond

Description: The site consists of three small, rocky islands (Little Island, Long Island, and Wantry Island) in western Oneida Lake. Little Island and Long Island are administered by the New York State Department of Transportation. Wantry Island is privately owned.

Birds: Common Terns have nested on islands in Oneida Lake since at least the 1930's. In 1982, approximately 400 pairs nested here; in 2003, there were 449 pairs. In 1997, there were 264 pairs of Double-crested Cormorants, 846 pairs of Ring-billed Gulls, 50 pairs of Herring Gulls, and one pair of Great Black-backed Gulls breeding at the site. At Sylvan/Verona Beach on the east end of the lake, there are migrating Bonaparte's Gulls, and post-breeding foraging sites for about 250 Common Terns. Areas of the lake that thaw in early spring can attract thousands of ducks. In mid- to late May, thousands of Brant and White-winged Scoters migrate along the lake early in the morning, having probably used the lake as a nighttime stopover location.

Conservation: A major issue in maintaining Common Tern breeding sites is preventing Ring-billed, Herring, and Great Black-backed Gulls from establishing nests in preferred tern nesting areas before the terns return in the spring. New York Cooperative Fish and Wildlife Research Unit staff have made efforts to prevent the gulls from nesting in certain areas, including installation of temporary monofilament gull exclusion grids, and gull nest removal under federal and state permits. In recent years, most successful Common Tern nesting activity has occurred on Little Island. However, starting in 2003, Common Terns have displayed a renewed interest in re-populating Wantry and Long Islands. Biologists with the New York Cooperative Fish and Wildlife Research Unit have worked to promote the expansion of tern nesting areas to these islands, using tern decoys and audio recordings. As most colonial nesting birds are sensitive to human disturbance and intrusion into nesting colonies, this should be prevented during the breeding season. Monitoring of Common Tern colonies should continue.

Common Tern

Onondaga Lake

Multiple municipalities, Onondaga County

3,000 acres
350-390' elevation

43.0897°N
76.2080°W

IBA Criteria Met

Criterion	Species	Data	Season	Source
Congregations-Waterfowl	Mixed species	3,639 ind. in 2003, 1,143 in 2002, 2,534 in 2001, 5,255 in 2000, 2,079 in 1999, 1,806 in 1998, 1,945 in 1997, 1,811 in 1996, 7,443 in 1995, 1,248 in 1994, 2,768 in 1993	Winter	Christmas Bird Count

Description: Approximately five miles long and one mile wide, Onondaga Lake is adjacent to Syracuse—one of the major urban areas in upstate New York—and is largely surrounded by industrial and urban development. The lake flows into the Oswego River, which empties north into Lake Ontario at Oswego Harbor. The lake, long used for dumping industrial wastes and sewage from Syracuse and other communities, is one of the most heavily polluted bodies of water in the state. Onondaga Lake and its tributaries are large contributors of sediment and associated nutrients to Lake Ontario.

Birds: This site is an important waterfowl wintering area. The most abundant species is the Common Merganser (4,650 in 2000; 6,200 in 1995). Bald Eagles (two to four individuals) are common in winter.

Conservation: Onondaga Lake is heavily polluted due to years of industrial dumping, chemical waste beds, and storm sewage runoff from the city of Syracuse's sewage treatment facility. Commercial development along the lakeshore is expanding as industrial use declines. The site is a recipient of New York State Environmental Bond Act funds, as well as federal funds for water pollution cleanup. The SUNY College of Environmental Science and Forestry has initiated an education program to develop tools for environmental monitoring, and for communicating the importance of the Onondaga Lake Watershed. The Onondaga Lake Partnership (OLP) was established in 2000 to promote cooperation among government agencies and other parties involved in managing Onondaga Lake and its watershed. The partnership coordinates the development and implementation of improvement projects to restore, conserve, and manage the lake, in accordance with the Onondaga Lake Management Plan. Waterfowl populations should be monitored. Studies on contaminant loads in the tissues of waterfowl and the possible adverse effects of these contaminants are needed, particularly for Common Mergansers.

Pharsalia Woods

Multiple municipalities, Chenango County

23,000 acres
1,180-1,975' elevation

42.5631°N
75.7791°W

IBA Criteria Met

Criterion	*Species*	*Data*	*Season*	*Source*
Responsibility Species Assemblage-Forest	Northern Flicker, Least Flycatcher, Wood Thrush, Canada Warbler, Scarlet Tanager, Rose-breasted Grosbeak	Breed	Breeding	NY BBA 2000

Description:

This site includes a series of high elevation forests surrounded by open farmland. The area is mostly hardwood forest (maple, beech, hemlock), and in swampy areas, red maple and red spruce stands. According to the NY GAP land cover data, approximately 85% of the site is forested, which includes deciduous wetland, evergreen northern hardwood, evergreen plantation, successional hardwood, and sugar maple mesic forests. State Route 23 divides the site, with the southern portion including the New Michigan State Forest, and the northern portion including the Pharsalia Wildlife Management Area—the first WMA established in New York. Much of the land is abandoned farmland that was purchased by the state in the 1930's. The site is popular for hunting, hiking, skiing, and snowmobiling.

Birds: Only the Catskills, the Adirondacks, Tug Hill Plateau, and Allegany State Park are higher in elevation than Pharsalia Woods, the fifth highest spot in the state and one of the few locations outside of these areas with breeding Swainson's Thrushes. The site supports an abundance of breeding forest birds. Characteristic breeders include the Northern Harrier, Sharp-shinned Hawk, Cooper's Hawk, Northern Goshawk, Red-shouldered Hawk, Eastern Wood-Pewee, Red-eyed Vireo, Veery, Swainson's Thrush, Hermit Thrush, Magnolia Warbler, Black-throated Blue Warbler, Yellow-rumped Warbler, Black-throated Green Warbler, Blackburnian Warbler, Mourning Warbler, Canada Warbler, Scarlet Tanager, and Rose-breasted Grosbeak. This site has some of the largest unfragmented blocks of hardwood and mixed forest in western New York. The area also has significant red pine and Norway spruce plantations; the seeds of these trees support winter finches such as the Pine Grosbeak, Red Crossbill, White-winged Crossbill, Common Redpoll, and Pine Siskin during irruption (high visitation) years.

Conservation: The NYS DEC manages the site for multiple uses. Portions of this site have been designated as a state Bird Conservation Area. The area is in great need of bird inventory and monitoring.

Queen Catharine Marsh

Multiple municipalities, Schuyler County

1,100 acres
455-565' elevation

42.3666°N
76.8539°W

IBA Criteria Met

Criterion	*Species*	*Data*	*Season*	*Source*
Species at Risk	Least Bittern	Estimated 8 pairs over the past 10 years	Breeding	John Gregoire pers. comm. 2004

Description: This site is a large cattail marsh at the southern end of Seneca Lake, between Watkins Glen and Montour Falls. Portion of the marsh are administered by the NYS DEC (Catharine Creek Marsh Wildlife Management Area) and the remainder is privately owned. This is the last remaining headwater marsh in the Finger Lakes and covers over 1,000 acres. The area, named for a past local Seneca Indian queen, Catharine Montour, provides a haven for many wildlife species. Once navigable to Montour Falls, the waters of Catharine Creek still feed a remnant section of the Chemung Barge Canal, which runs through the center of the marsh. The U.S. Army Corps of Engineers' flood control projects now funnel most of the water through a canal system. A few hillside creeks and John's Creek continue to feed the marsh.

Birds: This area supports at-risk species, including the American Black Duck (migrant), Common Loon (migrant), Pied-billed Grebe (confirmed breeder), American Bittern, Least Bittern, Osprey (first nest seen in 2003), Bald Eagle (migrant), American Woodcock, Willow Flycatcher, Sedge Wrens (sporadic), Wood Thrush, Blue-winged Warbler, Prothonotary Warbler (attempted nesting), and Rusty Blackbird (migrant). Other wetland-dependent species that breed here include Virginia Rails, Soras, Marsh Wrens, and Swamp Sparrows.

Conservation: This site is listed in the 2002 State Open Space Conservation Plan as a priority site under the project name Catharine Valley Complex. Potential development along Route 14/414 and Rock Cabin Road threaten the site. Attempts to protect the road and riparian area have been ineffective. Wildlife and avian activity and status are monitored, with reports issued by the Kestrel Haven Avian Migration Observatory.

© Jillian Liner

Queen Catharine Marsh in spring

Ripley Hawk Watch

Ripley and Westfield, Chautauqua County

16,000 acres
565-1,680' elevation

42.2597°N
79.6380°W

IBA Criteria Met

Criterion	*Species*	*Data*	*Season*	*Source*
Congregations-Raptors	Mixed species	Spring totals: 19,298 ind. in 2003, 11,133 in 2002, 12,603 in 2001, 22,422 in 2000, 14,002 in 1999, 22,603 in 1998, 14,652 in 1997, 14,287 in 1996, 12,530 in 1995, 9,439 in 1994, 13,056 in 1993, 23,688 in 1992	Spring	The Ripley Hawk Watch

Description: This site is located on a small ridge near the shore of Lake Erie in the southwest corner of New York State, approximately 4.5 miles southwest of the village of Westfield on Lake Erie. It is a mosaic of generally lowland forests, pastures, agricultural fields, and vineyards. The site is largely privately owned with multiple landowners.

Birds: The site is a major spring hawk concentration area. The annual number of raptors counted each spring (1992-2003) has varied from a low of 9,426 (1994) to a high of 23,688 (1992). The average seasonal total of hawks from 1992-2003 was 15,801. In 2003, a record 86 Bald Eagles and 20 Golden Eagles were documented. This site has reported the highest number of Sandhill Cranes on any migration flyway in the northeast (33 in 2003, 59 in 2004). Rare birds observed during migration have included the Black Vulture, Swallow-tailed Kite, Mississippi Kite, Swainson's Hawk, and Ferruginous Hawk (the first ever recorded in the state in 2002). This site also supports breeding species at risk, including the Bald Eagle (two pairs in the vicinity), Northern Harrier (two pairs), Sharp-shinned Hawk (two pairs), Cooper's Hawk (one pair), Red-shouldered Hawk (two pairs), and Red-headed Woodpecker (confirmed breeders). In recent years, the area has become a wintering site for Bald Eagles that use openings on Lake Erie. Recent studies indicate that this site is also an important migration corridor for nocturnal passerines.

Conservation: Regular and more comprehensive monitoring of spring hawk numbers should continue. The surrounding areas are likely important foraging and roosting habitat for migrating and nesting raptors; better documentation on how raptors use the area is needed. Also, additional monitoring of night-migrating passerines is recommended. A recently proposed wind power development has sparked concern among local birders, state conservation groups, and government agencies. Wind power facilities have the potential to negatively affect birds and other wildlife through collisions and habitat degradation from turbine construction and maintenance. Proper location of wind farms can help avoid such impacts. The progress of this proposal should be monitored closely.

Rochester Area Urban Parks

Multiple Municipalities, Monroe County

1,000 acres
275-590' elevation

43.2231°N
77.5746°W

IBA Criteria Met

Criterion	*Species*	*Data*	*Season*	*Source*
Congregations-Migrant landbirds	Mixed species	Hundreds of ind. (max. 845 in 1996) of over 50 species, including 20 warblers, can be observed during a few hours on fall or spring migration mornings	Migration	Nomination, Robert Marcotte

Description:

This site includes three parks located within the highly urbanized Rochester area. Cobbs Hill Park is located within the limits of the city of Rochester. Badgerow Park (formerly Dewy-Latta Park) and Durand-Eastman Park are outside the city limits. Durand-Eastman, the largest park (794 acres), includes a mile-long stretch of sand and rock beach on Lake Ontario. Cobbs Hill Park (125 acres) is located about six miles from Lake Ontario on one of the few hills in Rochester. A water supply reservoir lies within this park, which is surrounded by a densely populated urban area. Badgerow Park (66 acres) is located within the urban Rochester area, about one mile from Lake Ontario. The parks are owned by the city of Rochester (Cobbs Hill), the town of Greece (Badgerow), and Monroe County (Durand-Eastman).

Birds:

Together these parks serve as islands of forest habitat amidst a sea of urban development, and support an exceptional number and diversity of migrant passerines. The three parks are particularly important as migrant stopover locations. Regular surveys have documented hundreds (and sometimes thousands) of warblers and other songbirds using these areas during the spring and fall migration. In Badgerow Park alone, over 200 individuals were documented over a few hours on May 10, 1996, and 61 species, including 20 warblers, were recorded over a few hours on May 18, 1996. Total numbers of individuals observed during fall migration include 477 in 1995, 845 in 1996, 700 in 1997, and 682 in 1998. In a single day, 118 individuals of 24 species have been recorded at Cobbs Hill. In Badgerow Park, 80 Nashville Warblers, 215 Magnolia Warblers, 64 Yellow-rumped Warblers, 101 Black-throated Green Warblers, 70 Blackburnian Warblers, and 49

Wilson's Warblers were tallied in the fall of 1996. At same site during May 1996, 120 Nashville Warblers, 54 Magnolia Warblers, 56 Black-throated Blue Warblers, 75 Yellow-rumped Warblers, and 63 Ovenbirds were tallied. Maximum one-day warbler counts at Badgerow Park during September 1996 were 52 and 45. Maximum one-day spring counts during May 1996 were 116 and 90. Cobbs Hill Park's maximum one-day spring warbler count was 118 in May 1996 and the maximum fall count was 52 in fall 1996. Other maximum one-day warbler counts at Cobbs Hill were 15 Magnolia Warblers (spring 1996), 15 Black-throated Blue Warblers, 75 Yellow-rumped Warblers (fall 1996), 50 Bay-breasted Warblers (fall 1996), 15 American Redstarts (spring 1996), and 15 Ovenbirds (spring 1996).

Yellow-rumped Warbler

Conservation: All three parks are managed primarily as recreation areas and there is concern that wooded areas used extensively by birds may be susceptible to more intensive development for recreation, or increased recreational use. Park management should consider maintaining natural areas for migratory birds. Clearing of underbrush to "neaten" these areas is often detrimental to migratory birds. Durand-Eastman Park has a deer overpopulation problem, which decreases the amount of available shrubby understory habitat in some areas. Monitoring of migratory species at these sites should continue, and should include banding if possible.

Salmon Creek

Lansing, Tompkins County

500 acres
590-855' elevation

42.6060°N
76.5388°W

IBA Criteria Met

Criterion	*Species*	*Data*	*Season*	*Source*
Species at Risk	Cerulean Warbler	45 ind. in 1997, 11 in 1998; 6 in 2004	Breeding	Cerulean Warbler Atlas Project

Description: This forested site extends on either side of a one-mile stretch of Salmon Creek in the Finger Lakes region. The forest includes large sycamores, cottonwoods, black walnuts, and black locusts, with a thick understory of raspberries, blackberries, and other shrubs. There are several hemlock gullies along the smaller creeks that flow into Salmon Creek. Most of the land is privately owned, but the Finger Lakes Land Trust owns 33 acres.

Birds: The site has supported a major breeding concentration of at-risk Cerulean Warblers (estimated 46 pairs in 1997), along with a diversity of other songbirds, including Hooded Warblers, Yellow-bellied Sapsuckers, Acadian Flycatchers (three pairs in 1997), Yellow-throated Vireos, Blue-gray Gnatcatchers, Mourning Warblers, and Dark-eyed Juncos.

Conservation: Residential development along other portions of the creek has increased in recent years, and could affect the site if preservation efforts are not pursued. In some areas, large sycamores have been logged. Since large, mature trees are a necessary component of Cerulean Warbler breeding habitat, any loss of such trees could negatively impact their populations. The Finger Lakes Land Trust has purchased 33 acres in this area, and hopes to acquire more land or easements. The Cerulean Warbler population at the site has declined in recent years and targeted surveys should continue.

Seneca Army Depot

Romulus and Varik, Seneca County

10,000 acres
575-740' elevation

42.7422°N
76.8630°W

IBA Criteria Met

Criterion	*Species*	*Data*	*Season*	*Source*
Responsibility Species Assemblage-Shrub	American Woodcock, Willow Flycatcher, Brown Thrasher, Blue-winged Warbler, Eastern Towhee, Field Sparrow	Observed in breeding season	Breeding	Audubon NY statewide IBA field surveys

Description: This site encompasses approximately 10,000 acres between Cayuga and Seneca Lakes in the Finger Lakes region. Having formerly been maintained as a military base, much of the area is successional habitat. According to the NY GAP data, approximately 80% of the site is shrub habitat, which includes old field/pastures, shrub swamps, successional hardwoods, and successional shrubs. The site also has extensive forest and some wetland habitat, and supports a population of white deer.

Birds: This relatively large early successional habitat supports characteristic species, including the American Woodcock, Willow Flycatcher, Brown Thrasher, Blue-winged Warbler, Eastern Towhee, and Field Sparrow. A Red-headed Woodpecker was reported in 2003.

Conservation: This site is currently owned by the Seneca County Industrial Development Authority and ownership is expected to be transferred in the near future. Efforts to conserve the shrub/early-successional forest habitats at the site are strongly encouraged.

Seneca Lake

Multiple municipalities, Schuyler, Seneca, Ontario, and Yates Counties

42,700 acres
445-590' elevation

42.6519°N
76.8946°W

IBA Criteria Met

Criterion	*Species*	*Data*	*Season*	*Source*
Congregations-Waterfowl	Mixed species	8,162 ind. in 2004, 2,613 in 2003, 10,898 in 2002, 11,965 in 2001, 6,504 in 2000	Winter	NYSOA winter waterfowl count
Congregations-Individual Species	Pied-billed Grebe	Site supported 2% of state estimated winter population in 2000, 9% in 2001, 5% in 2002, 11% in 2003, 26% in 2004	Winter	NYSOA winter waterfowl count
Congregations-Individual Species	Mallard	Site supported 7% of state estimated winter population in 2000, 4% in 2001, 3% in 2002, 3% in 2003, 2% in 2004	Winter	NYSOA winter waterfowl count
Congregations-Individual Species	Redhead	Site supported 12% of state estimated winter population in 2000, 8% in 2001, 39% in 2004	Winter	NYSOA winter waterfowl count

Description: Seneca Lake is the second largest of New York's Finger Lakes. The lake spans four counties, with the city of Geneva on its northern end and Watkins Glen at its southern end. To the west lie Keuka and Canandaigua Lakes, and to the east, Cayuga Lake. It is approximately 32 miles long from north to south, but relatively narrow. Glacially carved in origin, it is the deepest of the Finger Lakes (620 feet). Much of the lakeshore is developed or in agriculture, but there are scattered marshes and wetlands. The lake itself is owned by the state of New York, but the lakeshore includes mostly private land, with some municipal and state-owned land.

Birds: The lake is an important waterfowl wintering area; surveys administered by the New York State Ornithological Association (NYSOA) have documented thousands of individuals using the lake each winter. During 2000-2004, the lake supported an average of eight Pied-billed Grebes (6% of state wintering population), 1,686 Mallards (4% of state wintering population), 1,737 Redheads (11% of state wintering population), 183 scaup, and 140 Common Goldeneyes.

Redhead

Counts of Canada Geese, American Black Duck, and Buffleheads have sometimes numbered in the hundreds during fall and winter.

Conservation: The introduction of zebra mussels and non-native fish may be having a negative effect on the aquatic ecosystem and the prey base of certain waterfowl species. However, some diving ducks feed extensively on zebra mussels and may benefit from their increased presence. Manipulation of water levels may impact waterfowl use of the lake. More research is needed to understand how different species of waterfowl are or will be impacted by changes in the lake ecosystem. The site is heavily used from spring through fall for recreational boating, fishing, and related activities. Disturbance does not seem to be a major problem since the largest concentrations of waterfowl occur before and after the peak of boating activity. Pollution from various sources, including non-point source agricultural runoff and boating activities, could impact the birds here and should be monitored. Waterfowl numbers should continue to be monitored.

Southern Skaneateles Lake Forest

Multiple municipalities, Onondaga and Cayuga Counties

9,500 acres
860-2,100' elevation

42.7684°N
76.2731°W

IBA Criteria Met

Criterion	*Species*	*Data*	*Season*	*Source*
Species at Risk	Cerulean Warbler	7+ pairs	Breeding	Matt Young pers. comm. 2004

Description: This site consists of the forested habitat at the southern end of Skaneateles Lake, including the state-owned Bear Swamp State Forest and the Finger Lakes Land Trust's High Vista Preserve. Skaneateles Lake, the steep valley walls, and the flattop ridges are all the result of glaciation, which occurred about 10,000 years ago. Historically, much of the area was cleared for farms. Today pine, spruce, hemlock, and hardwood forests dominate the site. There is a wetland complex at the southern end of the lake.

Barred Owl

Birds: At-risk species supported at this site include the American Bittern, Northern Harrier (at least one pair), Sharp-shinned Hawk, Cooper's Hawk (one pair), Northern Goshawk (one pair), Red-shouldered Hawk (one pair), Broad-winged Hawk (at least one pair), American Woodcock, Wood Thrush, Golden-winged Warbler, Cerulean Warbler, and Canada Warbler. Other notable species include the Virginia Rail, Wilson's Snipe, Acadian Flycatcher (locally rare), Common Raven, Veery, Hermit Thrush, White-throated Sparrow (locally rare), crossbill, and Pine Siskin. There are 20 plus breeding species of warblers at this site, including the Blue-winged Warbler, Nashville Warbler, Black-throated Blue Warbler, Blackburnian Warbler, Pine Warbler, Cerulean Warbler (seven plus pairs), Northern and Louisiana Waterthrush, Mourning Warbler, Hooded Warbler, and Canada Warbler.

Conservation: Some state owned lands are managed for timber, and sustainable forestry and best management practices should be promoted. Cerulean Warbler habitat requirements should be considered explicitly. Interest in acquiring land for conservation purposes is strong, due to the low level of development and botanical importance of this area. Each year, a nesting pair of Northern Goshawks is monitored and the nestlings are banded.

The Center at Horseheads Fields

Horseheads, Chemung County

175 acres
890-945' elevation

42.1861°N
76.8201°W

IBA Criteria Met

Criterion	*Species*	*Data*	*Season*	*Source*
Species at Risk	Grasshopper Sparrow	1- 44 ind. in breeding season last 6 years, multiple sightings each year	Breeding	Bill Ostrander pers. comm. 2004

Description: This site consists of a long-fallow field with an old airstrip (in poor condition), which serves as an access road. The area is the only undeveloped industrial-zoned property in the town. Development around the fringe of the field has occurred in recent years, while the center of the field is mowed for the use of radio-controlled airplanes. Several buildings have been erected along the northern border. There is an Army depot and a large private building at the south end. Several acres in the northern portion have been developed into athletic fields. The site is under a mix of private, corporate, municipal, and federal ownership.

Birds: This site has been an important grassland bird breeding area since 1968. Breeding species include the Northern Harrier (one pair in 2000), American Kestrel (one pair in 1996, 1998, and 2000), Upland Sandpiper (four in 1996, one in 1999, one in 2001), Horned Lark (seven in 1997, two in 1998, five in 1999, four in 2000, four in 2001, three in 2004), Savannah Sparrow (20 in 1996, 44 in 1999, 15 in 2000, 12 in 2001, 16 in 2002), Grasshopper Sparrow (six in 1996, 26 in 1999, 10 in 2000, 12 in 2001, eight in 2002, three in 2004), and Eastern Meadowlark (eight plus in 1996, five in 1997, 16 in 1998, 14 in 1999, 13 in 2000). Henslow's Sparrows have been present in spring (one in 1996), and in summer (one in 2001). Vesper Sparrows have been present in spring (one in 2000, one in 2004). This is the only historic nesting site of Upland Sandpipers in the southern Finger Lakes region. During periods of heavy rain, migrating shorebirds can be found in the puddles that form on the site; Greater and Lesser Yellowlegs are the most frequent visitors, but Black-bellied Plovers, American Golden-Plovers, Whimbrels, and Wilson's Snipe have also been recorded. The site occasionally hosts wintering raptors, including Rough-legged Hawks and Snowy Owls (one in 1993).

Conservation: The construction of athletic fields has already reduced the size of the area available for grassland birds, and planned increases in commercial development at the site will further reduce available habitat. Building construction in recent years has covered at least one known Upland Sandpiper nesting site. No specific conservation measures have been taken, except to notify the Town of Horseheads that several state-listed species occur at the site. Monitoring of grassland species should continue and a conservation management plan should be developed that provides for the needs of these birds.

© Jerry and Sherry Liguori

American Kestrel

Tifft Nature Preserve

Buffalo, Erie County

275 acres
565-580' elevation

42.8428°N
78.8559°W

IBA Criteria Met

Criterion	*Species*	*Data*	*Season*	*Source*
Species at Risk	Least Bittern	3 pairs over the past several years	Breeding	Karen Wallace pers. comm. 2004

Description: This site, once a landfill, is located within the city of Buffalo along the southern edge of the city, between a major rail yard and the Buffalo Outer Harbor. It includes a 75-acre cattail marsh (the second largest in Erie County) with open water ponds, a 50-acre upland mound with grasslands, and two large ponds in the west and northwest (former canals). The site is owned by city of Buffalo and managed by the Buffalo Museum of Science.

Birds: A total of 265 bird species have been recorded at this site, representing exceptional diversity for a site of this size. Sixty-six of these species have been found breeding or were present during the breeding season. At-risk species that regularly breed on the preserve include the Pied-billed Grebe and Least Bittern. The American Bittern occurs regularly and has bred irregularly. At-risk species that visit regularly include the Common Loon, Osprey, Bald Eagle, Northern Harrier, Red-shouldered Hawk, Peregrine Falcon, Upland Sandpiper, Common Tern, Black Tern (formerly bred), Short-eared Owl, Willow Flycatcher, and Sedge Wren (may have bred). The site is an important stopover site for migrants because of its proximity to Lake Erie, and because it is largely separated from other green space by industrial and commercial developments.

Conservation: This site is listed in the 2002 State Open Space Conservation Plan as a priority site under the project name Urban Wetlands. The focus of Tifft Nature Preserve is research and education. The encroachment of common reed (*Phragmites australis*) and purple loosestrife (*Lythrum salicaria*) into the wetlands may decrease habitat value for wetland birds. Biological control measures for purple loosestrife have begun. Pollution, possible future development near the site, and overuse of sensitive natural areas are all potential problems that should be monitored. Continued evaluation of measures to ensure biological integrity is necessary. Inventory and monitoring of birds, especially at-risk species, should continue. In 1999, a marsh restoration project that involved creating a 9,200 foot long channel, 20-24 feet in width, and approximately five feet deep, was completed with funding from an EPA Habitat Restoration Grant and wetland mitigation funds. This improvement resulted in better nesting habitat. Additional funds from the Clean Water Bond Act and USFWS will enable continued restoration to attain a hemi-marsh condition, with the goal of restoring Black Tern habitat. A new electric pump was acquired with funding from Ducks Unlimited to maintain proper water levels. Aquatic field testing has shown an improvement in water quality and decreased pollution. During the first round of IBA site identifications, this site was recognized under the research criterion because a long-term research project is based there.

Tifft Nature Preserve

Tioughnioga River/ Whitney Point Reservoir

Multiple municipalities, Broome and Cortland Counties

17,600 acres
885-1,480' elevation

42.4099°N
75.9564°W

IBA Criteria Met

Criterion	*Species*	*Data*	*Season*	*Source*
Responsibility Species Assemblage-Shrub	American Woodcock, Gray Catbird, Brown Thrasher, Blue-winged Warbler, Prairie Warbler, Eastern Towhee, Field Sparrow, Indigo Bunting	Breed [1] and [2]	Breeding	1. NY BBA 2000; 2. Audubon NY statewide IBA field surveys

Prairie Warbler

Description: The Whitney Point Reservoir was created in 1942 by the construction of a flood control dam across the Otselic River. The dam and water levels are maintained by the U.S. Army Corps of Engineers. From 1942 to 1964 no permanent body of water was impounded behind the dam, but in 1964, a 1,200-acre, 4.5 mile recreational lake was created, extending from Whitney Point to Upper Lisle. Two three-acre ponds with emergent wetlands were also created. These ponds, along with the Otselic River backwaters, now constitute the Dorchester Park/ Whitney Point Multiple Use Area. According to the NY GAP land cover data, approximately 40% of the site is shrub habitat, which includes old field/pastures, shrub swamps, successional hardwoods, and successional shrubs. The site includes county, state, and federally owned lands.

Birds: This site is an important breeding location for a suite of shrub bird species, including American Woodcock, Gray Catbird, Brown Thrasher, Blue-winged Warbler, Prairie Warbler, Eastern Towhee, Field Sparrow, and Indigo Bunting. Due to its location in a portion of the state with few large bodies of water, the reservoir is also an important waterfowl stopover location, hosting a regionally high abundance and diversity of waterfowl. In 1997, the site supported 4,000 Snow Geese, Brant, 12 Tundra Swans, Gadwalls, American Wigeons, American Black Ducks, Blue-winged Teal, Northern Shovelers, Northern Pintails, Green-winged Teal, Canvasbacks, Ring-necked Ducks, Greater Scaup, Lesser Scaup, Black Scoters, Long-tailed Ducks, Buffleheads, Common Goldeneyes, 100 Hooded Mergansers, 484 Common Mergansers, and Ruddy Ducks. The site regularly hosts raptors, including Ospreys and Bald Eagles. In years when the water is drawn down, the exposed mudflats support a variety of shorebirds. For example, in 1992, birders noted the Black-bellied Plover, Semipalmated Plover, Lesser Yellowlegs, Hudsonian Godwit, White-rumped Sandpiper, Pectoral Sandpiper, and Dunlin at the site. The Upper Lisle woods section is an excellent songbird migration stopover location in May, averaging 16-18 species of warblers, thrushes, vireos, and others.

Conservation: Approximately 15 acres on the western side of the impoundment and 25 plus acres at the north end are being managed to restore grassland habitat. A more in-depth inventory of site usage by birds is needed.

Toad Harbor Swamp

Constantia and West Monroe, Oswego County

3,600 acres
365-395' elevation

43.2569°N
76.0752°W

IBA Criteria Met

Criterion	*Species*	*Data*	*Season*	*Source*
Species at Risk	Cerulean Warbler	7 ind. in 1997 [1]; estimate of 10+ pairs in 2004 [2]	Breeding	1. Cerulean Warbler Atlas Project; 2. Bill Purcell pers. comm. 2005
Congregations-Waterfowl	Wood Duck	2,590 ind. in 1996	Migration	Nomination, Bill Purcell and pers. comm. 2004
Congregations-Wading Birds	Great Blue Heron	38 nests in 1996, 50+ nests in 2004	Breeding	Bill Purcell pers. comm. 2004

Description: This site is administered by the NYS DEC and includes a hardwood swamp with beaver ponds, upland woods, fields, brush, and marsh habitats.

Birds: The mix of habitats support many at-risk species, including the Pied-billed Grebe (four plus pairs), American Bittern (breeder), Least Bittern (present and presumed breeding), Osprey (bred in 2001, probably still breeding), Northern Harrier (one pair), Sharp-shinned Hawk (present), Cooper's Hawk (breeds), Red-shouldered Hawk (breeds), American Woodcock (common), Common Nighthawk (present), Red-headed Woodpecker (confirmed breeding in late 1990's), Willow Flycatcher (common), Sedge Wren (sings in some years but doesn't persist), Wood Thrush (common), Blue-winged Warbler (present), Golden-winged Warbler (historically present), Cerulean Warbler (breeds, seven individuals in 1997, an estimated 10 plus pairs in 2004), Prothonotary Warbler (at least two pairs), Grasshopper Sparrow (present in 2004), and Henslow's Sparrow (not found since late 1990s). The area also has a Great Blue Heron rookery (50 plus nests), and breeding Green Herons, Virginia Rails, Soras, Common Moorhens (at least three pairs), and a significant fall concentration of Wood Ducks (2,590 in 1996). The amount of Wood Duck habitat has increased in recent years and the numbers of individuals using this site may have also increased.

Cerulean Warbler

Conservation: This site is listed in the 2002 State Open Space Conservation Plan as a priority site under the project name North Shore of Oneida Lake. Local residents removed a beaver dam in one area, which resulted in a decreased amount of active wetlands. The fields should be mowed to maintain grassland bird habitat. Cerulean Warblers use very mature white oaks, yet the forest understory consists primarily of maple trees. Inventory and monitoring of at-risk species should continue.

Wheeler's Gulf

Pomfret, Chautauqua County

210 acres
1,010-1,150' elevation

42.4016°N
79.3332°W

IBA Criteria Met

Criterion	*Species*	*Data*	*Season*	*Source*
Species at Risk	Cerulean Warbler	Historically 4-5 pairs, but none heard in recent years	Breeding	Terrence Mosher pers. comm. 2005

Description: This site includes mature forests on both sides of a deep valley with a beaver pond complex. On the south-facing slope is a beech-hemlock forest, while the north-facing slope is predominately an oak-hickory forest. An abandoned railroad right-of-way extends through the center of the site along the valley floor. The site is mostly privately owned, but a central portion is owned by the Pennsylvania Railroad. The area is used for hiking, birding, and hunting.

Birds: This site supports an unusual diversity of breeding birds for the region, including the Acadian Flycatcher (historically four to six pairs, five singing males in 2004), Cerulean Warbler (historically one to four pairs), Hooded Warbler (present), Scarlet Tanager, Rose-breasted Grosbeak, and American Redstart. This is the only known breeding site for Cerulean Warblers in Chautauqua County.

Conservation: None of this site is currently under conservation protection. Efforts should be made to acquire conservation easements or fee titles for the land, including the railroad right-of-way. The site needs a more thorough inventory of at-risk species to better understand bird use and determine whether this site continues to meet the IBA criteria.

Whiskey Hollow

Van Buren, Onondaga County

600 acres
455-610' elevation

43.1166°N
76.3931°W

IBA Criteria Met

Criterion	*Species*	*Data*	*Season*	*Source*
Species at Risk	Cerulean Warbler	8 ind. in 1997; 7 singing males in 2004	Breeding	Bernie Carr pers. comm. 2004

Description: This site is located just outside of Syracuse, and includes a heavily wooded valley surrounded by farmland.

Birds: This site supports an unusual diversity of breeding birds, including the Sharp-shinned Hawk (observed during breeding season), Cooper's Hawk (two pairs in 2004), Black-billed Cuckoo (breeds), Yellow-billed Cuckoo (breeds), Barred Owls, Acadian Flycatcher (breeds), Yellow-throated Vireo (breeds), Winter Wren, Golden-winged Warbler (at least two pairs in 1996, but not seen in recent years), Northern Parula (one of few breeding sites in the area, but not seen in recent years), Cerulean Warbler (four to seven singing males, females and young observed in 2004), Mourning Warbler (breeds), Hooded Warbler (breeds), Scarlet Tanager (breeds, six singing males), and Vesper Sparrow (one pair in 1996).

Conservation: The site is mostly privately owned, although Save the County Land Trust owns 43 acres. Extensive logging of oak and maple trees is going on in some areas. The site should have an active plan for preservation and management to benefit the diverse breeding bird community, particularly at-risk Cerulean Warblers. Regular inventory and monitoring, especially for at-risk species, should continue.

Widger Hill

Caton, Steuben County

460 acres
1,580-1,780' elevation

42.0101°N
76.9718°W

IBA Criteria Met

Criterion	*Species*	*Data*	*Season*	*Source*
Species at Risk	Grasshopper Sparrow	6+ pairs in 1996	Breeding	Nomination, Bill Ostrander
Species at Risk	Henslow's Sparrow	6 pairs in 1996	Breeding	Nomination, Bill Ostrander

Description: This privately owned area includes grassy meadows interspersed with goldenrod. The hills provide long-distance viewing—a common feature of Henslow's Sparrow habitat. A road divides the site into northern and southern parts.

Birds: This is an important grassland bird breeding area, hosting the Horned Lark (one in 1998), Savannah Sparrow, Grasshopper Sparrow (at least six pairs in 1996), Henslow's Sparrow (six pairs in 1996), Bobolink (at least four pairs in 1996), and Eastern Meadowlark (six pairs in 1996).

Conservation: Vegetative succession is not an immediate concern, but the landowner should be encouraged to maintain the grassland. Owners of surrounding grasslands should also be encouraged to manage the habitat for grassland birds. Inventory and monitoring, particularly of at-risk species, is needed to better understand bird use, and to determine whether this site continues to meet IBA criteria.

Chapter 3.2
Northern New York

View of Adirondack Peaks

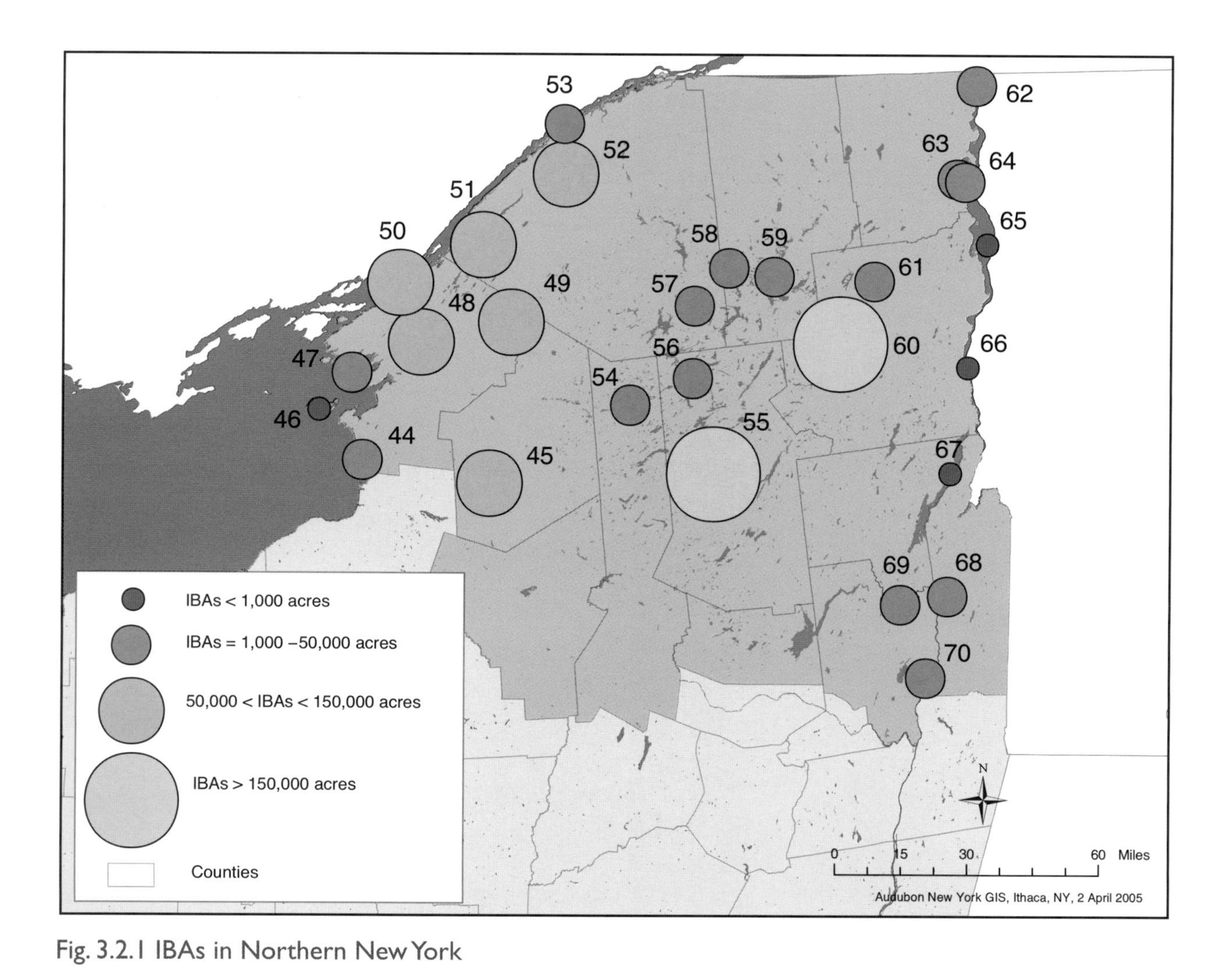

Fig. 3.2.1 IBAs in Northern New York

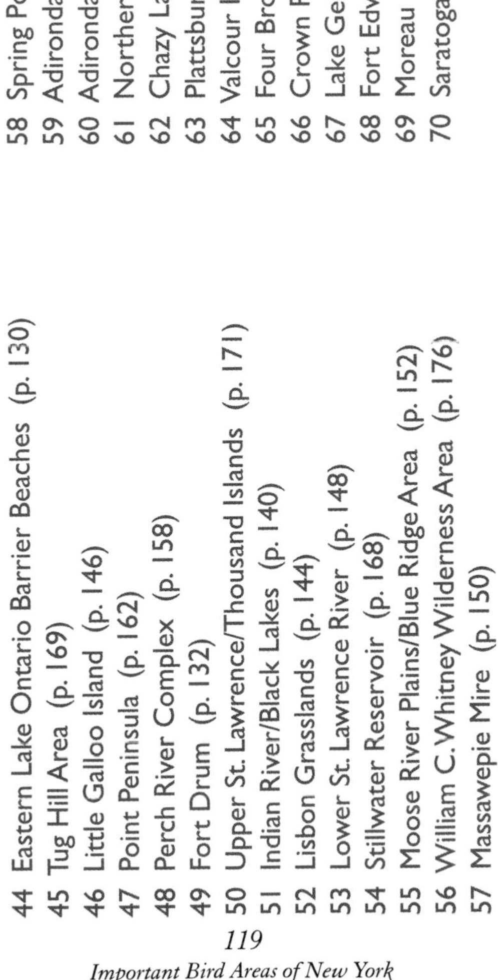

44 Eastern Lake Ontario Barrier Beaches (p. 130)
45 Tug Hill Area (p. 169)
46 Little Galloo Island (p. 146)
47 Point Peninsula (p. 162)
48 Perch River Complex (p. 158)
49 Fort Drum (p. 132)
50 Upper St. Lawrence/Thousand Islands (p. 171)
51 Indian River/Black Lakes (p. 140)
52 Lisbon Grasslands (p. 144)
53 Lower St. Lawrence River (p. 148)
54 Stillwater Reservoir (p. 168)
55 Moose River Plains/Blue Ridge Area (p. 152)
56 William C. Whitney Wilderness Area (p. 176)
57 Massawepie Mire (p. 150)
58 Spring Pond Bog Preserve (p. 166)
59 Adirondack Loon Complex (p. 124)
60 Adirondack High Peaks Forest Tract (p. 120)
61 Northern Adirondack Peaks (p. 157)
62 Chazy Landing/Kings Bay Area (p. 126)
63 Plattsburgh Airfield (p. 160)
64 Valcour Island (p. 174)
65 Four Brother Islands (p. 138)
66 Crown Point State Historic Site (p. 128)
67 Lake George Peregrine Site (p. 142)
68 Fort Edward Grasslands (p. 136)
69 Moreau Lake Forest (p. 156)
70 Saratoga National Historical Park (p. 164)

Adirondack High Peaks Forest Tract

Multiple municipalities, Essex, Franklin, and Hamilton Counties

353,000 acres
960–5,344' elevation

44.1080°N
74.0354°W

IBA Criteria Met

Criterion	*Species*	*Data*	*Season*	*Source*
Species at Risk	American Black Duck	Extent of habitat and breeding atlas presence strongly suggest that the threshold is being met [1]; 4CO, 2PR, 1PO [2]	Breeding	1. Technical Committee consensus; 2. NY BBA 2000
Species at Risk	Common Loon	4CO, 6PR, 9PO	Breeding	NY BBA 2000
Species at Risk	American Bittern	Extent of habitat and breeding atlas presence strongly suggest that the threshold is being met [1]; 1CO, 0PR, 0PO [2]	Breeding	1. Technical Committee consensus; 2. NY BBA 2000
Species at Risk	Osprey	Extent of habitat and breeding atlas presence strongly suggest that the threshold is being met [1]; 1CO, 1PR, 3PO [2]	Breeding	1. Technical Committee consensus; 2. NY BBA 2000
Species at Risk	Sharp-shinned Hawk	Extent of habitat and breeding atlas presence strongly suggest that the threshold is being met [1]; 2CO, 1PR, 1PO [2]	Breeding	1. Technical Committee consensus; 2. NY BBA 2000
Species at Risk	Cooper's Hawk	Extent of habitat and breeding atlas presence strongly suggest that the threshold is being met [1]; 2CO, 0PR, 1PO [2]	Breeding	1. Technical Committee consensus; 2. NY BBA 2000
Species at Risk	Northern Goshawk	Extent of habitat and breeding atlas presence strongly suggest that the threshold is being met [1]; 1CO, 0PR, 2PO [2]	Breeding	1. Technical Committee consensus; 2. NY BBA 2000
Species at Risk	Red-shouldered Hawk	Extent of habitat and breeding atlas presence strongly suggest that the threshold is being met [1]; 0CO, 0PR, 1PO [2]	Breeding	1. Technical Committee consensus; 2. NY BBA 2000
Species at Risk	Peregrine Falcon	3 pairs in 2002, 2 pairs in 2001, 4 pairs in 2000, 3 pairs in 1999-1997, 2 pairs in 1996-1995	Breeding	NY Natural Heritage Biodiversity Databases

Criterion	*Species*	*Data*	*Season*	*Source*
Species at Risk	American Woodcock	Extent of habitat and breeding atlas presence strongly suggest that the threshold is being met [1]; 0CO, 2PR, 2PO [2]	Breeding	1. Technical Committee consensus; 2. NY BBA 2000
Species at Risk	Olive-sided Flycatcher	Extent of habitat and breeding atlas presence strongly suggest that the threshold is being met [1]; 0CO, 4PR, 0PO [2]	Breeding	1. Technical Committee consensus; 2. NY BBA 2000
Species at Risk	Bicknell's Thrush	16 individuals recorded on 9 routes [1]; 1CO, 6PR, 7PO [2]	Breeding	1. Lambert, J.D. 2003; 2. NY BBA 2000
Species at Risk	Wood Thrush	Extent of habitat and breeding atlas presence strongly suggest that the threshold is being met [1]; 1CO, 2PR, 6PO [2]	Breeding	1. Technical Committee consensus; 2. NY BBA 2000
Species at Risk	Bay-breasted Warbler	Extent of habitat and breeding atlas presence strongly suggest that the threshold is being met [1]; 1CO, 1PR, 0PO [2]	Breeding	1. Technical Committee consensus; 2. NY BBA 2000

Bicknell's Thrush

Criterion	*Species*	*Data*	*Season*	*Source*
Species at Risk	Canada Warbler	Extent of habitat and breeding atlas presence strongly suggest that the threshold is being met [1]; 3CO, 6PR, 4PO [2]	Breeding	1. Technical Committee consensus; 2. NY BBA 2000
Species at Risk	Rusty Blackbird	0CO, 1PR, 0PO [1]; 5CO, 6PR, 5PO [2]	Breeding	1. NY BBA 2000; 2. NY BBA 1980
Responsibility Species Assemblage-Forest	Ruffed Grouse, American Woodcock, Yellow-bellied Sapsucker, Eastern Wood-Pewee, Least Flycatcher, Great Crested Flycatcher, Blue-headed Vireo, Veery, Bicknell's Thrush, Wood Thrush, Northern Parula, Chestnut-sided Warbler, Black-throated Blue Warbler, Black-throated Green Warbler, Blackburnian Warbler, Blackpoll Warbler, Black-and-white Warbler, American Redstart, Ovenbird, Canada Warbler, Scarlet Tanager, Rose-breasted Grosbeak, Purple Finch	Breed	Breeding	NY BBA 2000
Congregations-Individual Species	Bicknell's Thrush	16 individuals recorded on 9 routes	Breeding	Lambert, J.D. 2003

Description: This site is the largest relatively intact tract of forest habitat in the state. It also contains the largest legally designated Wilderness Area in the state. It is a mountainous region with a maximum elevation of 5,344 feet and at least 24 peaks over 3,500 feet, with sub-alpine and alpine habitats. According to the NY GAP land cover data, approximately 95% of the site is forested, which includes evergreen northern hardwood, deciduous wetland, evergreen plantation, spruce fir, evergreen wetland, and successional hardwood forests. The site also includes many lakes, ponds, rivers, and streams.

Birds: The peaks over 2,800 feet provide habitat for a distinctive sub-alpine bird community that includes the Bicknell's Thrush, Swainson's Thrush, and Blackpoll Warbler. Within the Wilderness Area is a vast amount of forest habitat that supports a characteristic forest breeding bird community, including the Ruffed Grouse, American Woodcock, Yellow-bellied Sapsucker, Eastern Wood-Pewee, Least Flycatcher, Great Crested Flycatcher, Blue-headed Vireo, Veery, Wood Thrush, Northern Parula, Chestnut-sided Warbler, Black-throated Blue Warbler, Black-throated Green Warbler, Blackburnian Warbler, Blackpoll Warbler, Black-and-white Warbler, American Redstart, Ovenbird, Canada Warbler, Scarlet Tanager, Rose-breasted Grosbeak, and Purple Finch. The boreal habitat supports the Spruce Grouse, Three-toed Woodpecker, Black-backed Woodpecker, Gray Jay, Boreal Chickadee, Red Crossbill, and White-winged Crossbill. The area also supports breeding at-risk species, including the Common Loon and Peregrine Falcon.

Conservation: Portions of this site have been designated as a state Bird Conservation Area and Wilderness Area. Significant restrictions apply to management activities on the state-owned portions of this area, which is one of the most popular outdoor recreation areas in the state. The number of visitors more than doubled from 1983 to 1993 to over 114,000 visitors per year. There is concern that this level of human use will degrade habitats, especially near trails and on fragile alpine habitats. More research is needed to examine the impacts of human visitation on bird populations and habitats. Opportunities exist for sustainable forest management on privately-owned portions of this site, which can provide required habitat for birds that prefer successional forests. Acid rain has had a major effect on the forest and lake ecosystems, but the long-term effects on birds like the Common Loon and Osprey are unclear. Acid rain may also be having an impact on the nesting success of songbirds, particularly at high elevations, by killing snails and other sources of calcium needed for egg production. The curtailment of sulphur dioxide emissions and the reduction of acid rain is currently a significant NY State initiative. Scientists in a number of academic institutions, government agencies, and non-government organizations operating in the park are considering launching an "All-Taxa Biological Inventory" of the park on public and private lands, to provide a better knowledge base for future management decisions. Detailed inventory and standardized monitoring of at-risk species is needed for the area. The Vermont Institute of Natural Science (VINS) coordinates the Mountain Birdwatch survey that monitors breeding Bicknell's Thrush and other high-elevation birds.

Adirondack Loon Complex

Multiple municipalities, Franklin, Hamilton, Herkimer, and St. Lawrence Counties

22,000 acres
1,475-1,805' elevation

44.0796°N
74.7815°W

IBA Criteria Met

Criterion	*Species*	*Data*	*Season*	*Source*
Species at Risk	Common Loon	160 adult loons and 13 chicks were observed in 2002	Breeding	Adirondack Cooperative Loon Program
Congregations-Individual Species	Common Loon	160 adult loons and 13 chicks were observed in 2002	Breeding	Adirondack Cooperative Loon Program

Common Loon

Description: This site is a complex of lakes in the Adirondack region that have each supported six or more adult loons in the breeding season. The lakes are Lows Lake, Cranberry Lake, Lake Clear, South Lake, Woodhull Lake, Honnedage Lake, Upper St. Regis Lake, Lake Bonaparte, Little Clear Pond, Little Tupper Lake, Canachagala Lake, Little Moose Lake, Bay Pond, Big Moose Lake, Little Long Pond, and Long Pond–Franklin City.

Birds: Each lake within this complex has supported over six Common Loons during the breeding season. The complex as a whole supports an exceptional number of breeding Common Loons.

Conservation: Potential factors impacting loon breeding success include lead poisoning, acid rain, shoreline development, human disturbance, and lake water levels. The Adirondack Cooperative Loon Program (ACLP), a collaborative research and education effort, conducts research to determine the status and trends in the Adirondack breeding loon population, the impact of mercury contamination on this population's reproductive success, and the migratory patterns and wintering areas of Adirondack loons. This work is coordinated with similar research throughout northeastern North America to determine the effect of mercury pollution and other factors on the breeding loon population throughout the region. ACLP also seeks to minimize human impacts on loon populations and other wildlife through a variety of public education projects.

Chazy Landing/Kings Bay Area

Champlain and Chazy, Clinton County

3,200 acres
95-105' elevation

44.8947°N
73.3833°W

IBA Criteria Met

Criterion	*Species*	*Data*	*Season*	*Source*
Species at Risk	Pied-billed Grebe	At least 6 breeding pairs every year	Breeding	Bill Krueger pers. comm. 2004
Species at Risk	Horned Lark	Hundreds can be seen in winter	Winter	Bob Budliger pers. comm. 2004
Responsibility Species Assemblage-Grassland	Killdeer, Upland Sandpiper, Bobolink, Eastern Meadowlark	Breeds	Breeding	NY BBA 2000

Description: This site is situated on the shore of Lake Champlain approximately 13 miles north of Plattsburgh, stretching from southwest of the outlet of the Little Chazy River north to the Kings Bay Wildlife Management Area. The site is mostly privately owned land, but the 421-acre Kings Bay Wildlife Management Area (WMA) is administered by the New York State Department of Environmental Conservation (NYS DEC). The Kings Bay WMA consists of hardwood swamps and cattail marshes. The remaining land is agricultural with many fallow fields. According to the NY GAP land cover data, approximately 60% of the site is open habitat, which includes cropland and old field/pasture. The lake and shore provide prime waterfowl habitat.

Birds: These wetlands and grasslands along Lake Champlain regularly support at-risk species including the American Black Duck (breeds most years), Common Loon (possible breeder), Pied-billed Grebe (at least six breeding pairs each year), American Bittern (regular breeder with two or more pairs), Osprey (individuals observed, potential breeder), Northern Harrier (year round resident, two pairs breeding), Peregrine Falcon (possible breeder), Upland Sandpiper (two breeding pairs each year from 1993-2002), Short-billed Dowitcher (regular migrant), Wilson's Phalarope (confirmed breeder in 1993, 1994, 1997, 2002), Common Tern (as many as 45 individuals observed during summer), Black Tern (up to eight individuals during breeding season), Short-eared Owl (up to eight individuals in 1995 winter), Willow Flycatcher (six singing males in 2001-2003), and Horned

Lark (abundant in late summer and winter, confirmed breeder). This is the first and only breeding site for Wilson's Phalarope in the state. Rare sightings of additional at-risk species include the Bald Eagle (occasional visitor), Sharp-shinned Hawk (rare migrant), Cooper's Hawk (winter), American Golden Plover (rare migrant), Whimbrel (rare migrant), Hudsonian Godwit (occasional migrant), Marbled Godwit (rare migrant), Red Knot (rare migrant), Sedge Wren (rare migrant), and Rusty Blackbird (rare migrant). The area also supports wintering Rough-legged Hawks, occasionally Snowy Owls, and, historically, Gray Partridges.

Conservation: This site is listed in the 2002 Open Space Conservation Plan as a priority site under the project name Lake Champlain Shoreline and Wetlands. Portions of this site have been designated as a state Bird Conservation Area. Much of the area is composed of farms that include cropland as well as pasture lands. Current land management practices have been beneficial to birds. Potholes and ditches have been created to hold water in an effort to improve the area for waterfowl and other wildlife. In addition, Wood Duck nest boxes have been erected throughout the area. Non-point source agricultural pollution is a potential problem for wetlands and should be monitored. Improved inventory and monitoring of at-risk species are needed.

Crown Point State Historic Site

Crown Point, Essex County

350 acres
90-95' elevation

44.0323°N
73.4238°W

IBA Criteria Met

Criterion	*Species*	*Data*	*Season*	*Source*
Congregations-Migrant Landbirds	Mixed species	Unusual diversity and abundance of migrants	Spring	John M.C. Peterson pers. comm. 2004

Description: This site includes a peninsula approximately two miles southeast of Port Henry on Lake Champlain, and is bordered on the east by Bulwagga Bay, on the north and east by Lake Champlain, and on the south by the hamlet of Burdick Crossing. This historic site is administered by New York State Office of Parks, Recreation and Historic Preservation (NYS OPRHP) and has numerous locations relating to the military history of the Lake Champlain and Lake George valleys. The area is a mixture of bottomland deciduous forests, wooded swamps, meadows, cedar/juniper scrub, and hawthorn groves. In additional to historical ruins, several modern buildings and parking lots are scattered around the tip of the peninsula.

Birds: Crown Point supports an unusual diversity and abundance of migratory songbirds; more than 180 species have been recorded. A spring banding station (sponsored by Crown Point Banding Association) has been in operation at the site since 1976; 12,036 birds of 95 species have been banded on Crown Point peninsula since 2004. At-risk species observed include the American Bittern (possible breeder), Osprey (probable breeder), Northern Harrier (probable breeder), Red-shouldered Hawk (possible breeder), American Woodcock (probable breeder), Whip-poor-will (possible breeder), Wood Thrush (confirmed breeder), Blue-winged Warbler (possible breeder), Golden-winged Warbler (possible breeder), and Prairie Warbler (possible breeder).

Conservation: This site is listed in the 2002 Open Space Conservation Plan as a priority site under the project name Lake Champlain Shoreline and Wetlands. A portion of this site has been approved for designation as a state Bird Conservation Area. Succession of the shrub/scrub habitat that supports the diversity and abundance of passerine migrants is taking place. Management plans for the site acknowledge the site's importance for migrant birds and will ensure maintenance of adequate amounts of suitable shrub/scrub habitat. Additional monitoring of at-risk species is needed to better understand bird use and determine if this site warrants IBA recognition under the at-risk criterion. During the first round of IBA site identifications, this site was recognized under the research criterion because it is a site where a long-term monitoring project is based.

Eastern Lake Ontario Barrier Beaches

Multiple municipalities, Oswego and Jefferson Counties

23,000 acres
240-395' elevation

43.6604°N
76.1534°W

IBA Criteria Met

Criterion	*Species*	*Data*	*Season*	*Source*
Species at Risk	Least Bittern	7 ind. in 2002; 2 pairs, 2 ind. in 2001	Breeding	NY Natural Heritage Biodiversity Databases
Species at Risk	Black Tern	18 ind. in 1998, 14 ind. in 2001	Breeding	Mazzocchi, I.M. and M. Roggie 2004
Congregations-Waterfowl	Mixed species flocks	Average 2,345, maximum 2,985 individuals from 2000-2003	Winter	NYSOA winter waterfowl counts
Congregations-Shorebirds	Mixed species flocks	Site regularly supports over 300 individuals [1]; average 664 maximum, 1,005 individuals from 1985-87 [2]	Fall	1. Gerry Smith pers. comm. 2004; 2. International Shorebird Surveys (ISS)

Description: This site extends roughly from Little Salmon River north to Black River and east (inland) to Route 3. It contains the remains of one of the largest inland dune systems in the Eastern Great Lakes and some of the highest quality freshwater marshes in the state. The site includes a mix of private and public land, with many significant wetlands administered by NYS OPRHP (Southwick Beach and Sandy Island Beach State Parks), NYS DEC (Deer Creek Marsh Wildlife Management Area [WMA]), Sandy Pond Beach Unique Area, Lakeview Marsh WMA and Black Pond WMA), and land owned by The Nature Conservancy (El Dorado Preserve).

Birds: This vitally important wetland complex supports many migrating and breeding species, including at-risk species such as the American Black Duck (winter), Common Loon (winter), Pied-billed Grebe (present in breeding season), American Bittern (present in breeding season), Least Bittern (breeds), Northern Harrier (breeds), Common Tern (has bred), Black Tern (breeds), and Sedge Wren (present in breeding season). Many wetland-dependent species use the area as a migratory staging and feeding area, including the Caspian Tern, Common Tern, various shorebird species, and a diversity of waterfowl. Numbers of all waterfowl fluctuate widely, depending on winter conditions. In mild winters the area hosts thousands of ducks, including the American Black Duck, Mallard, Long-tailed Duck, Common Goldeneye, and Common Merganser. As recently as 1984, the site supported breeding Piping Plovers.

Conservation: This site is listed in the 2002 Open Space Conservation Plan as a priority site under the project name Eastern Lake Ontario Shoreline and Islands. Portions of this site (Black Pond WMA, Lakeview Marsh WMA, Sandy Pond Beach Natural Area, and Deer Creek Marsh WMA), have been designated a state Bird Conservation Area. Much of the shoreline has been developed for camps and houses. Potential negative impacts of recreational use on sensitive species, particularly of sand beaches, are a concern and should be considered in future management. Approximately one half of the barrier dune is managed by NYS OPRHP, NYS DEC, and The Nature Conservancy. The New York Sea Grant program, The Nature Conservancy, NYS DEC, NYS OPRHP, and the Onondaga Audubon Society are working together to implement an educational outreach program with dune stewards to minimize and manage human use of the beach to prevent ecological damage. Invasive purple loosestrife (*Lythrum salicaria*) and common reed (*Phragmites australis*) threaten the marsh habitat and should continue to be eradicated. Water level fluctuations from flood control efforts may have a detrimental impact on native cattails (*Typha latifolia*). Inventory and monitoring of at-risk species should continue at the site. During the first round of IBA site identifications, this site was recognized under the research criterion because a long-term monitoring project is based there.

Fort Drum

Multiple municipalities, Jefferson, St. Lawrence, and Lewis Counties

107,000 acres
490-850' elevation

44.1116°N
75.5450°W

IBA Criteria Met

Criterion	*Species*	*Data*	*Season*	*Source*
Species at Risk	American Bittern	Recorded in 15 atlas blocks (3CO, 5PR, 7PO)	Breeding	NY BBA 2000
Species at Risk	Northern Harrier	Recorded in 8 atlas blocks (2CO, 4PR, 2PO) [1]; average 10-12 pairs each year [2]	Breeding	1. NY BBA 2000; 2. Mickey Scilingo pers. comm. 2004
Species at Risk	Red-shouldered Hawk	Recorded in 8 atlas blocks (2CO, 6PR)	Breeding	NY BBA 2000
Species at Risk	Upland Sandpiper	Recorded in 8 atlas blocks (1CO, 3PR, 4PO) [1]; at least 10 pairs in the mid-1990s, 4 pairs in 2004 [2]	Breeding	1. NY BBA 2000; 2. Mickey Scilingo pers. comm. 2004
Species at Risk	American Woodcock	Recorded in 17 atlas blocks (4CO, 9PR, 4PO)	Breeding	NY BBA 2000
Species at Risk	Common Nighthawk	Recorded in 9 atlas blocks (4CO, 4PR, 1PO) [1]; estimated 60-70 ind. in 2003 [2]	Breeding	1. NY BBA 2000; 2. Mickey Scilingo pers. comm. 2004
Species at Risk	Whip-poor-will	Recorded in 17 atlas blocks (1CO, 10PR, 6PO) [1]; 200+ ind. in 2003 [2]	Breeding	1. NY BBA 2000; 2. Mickey Scilingo pers. comm. 2004
Species at Risk	Sedge Wren	14-23 singing males in 2004	Breeding	Mickey Scilingo pers. comm. 2004
Species at Risk	Golden-winged Warbler	Recorded in 14 atlas blocks (4CO, 6PR, 4PO)	Breeding	NY BBA 2000
Species at Risk	Prairie Warbler	Estimated 25-30 singing males	Breeding	Mickey Scilingo pers. comm. 2004
Species at Risk	Canada Warbler	Recorded in 17 atlas blocks (8CO, 6PR, 3PO)	Breeding	NY BBA 2000
Species at Risk	Vesper Sparrow	Stable population, 100+ singing males in 2003	Breeding	Mickey Scilingo pers. comm. 2004

Criterion	*Species*	*Data*	*Season*	*Source*
Species at Risk	Grasshopper Sparrow	Stable population, 35+ singing males in 2003	Breeding	Mickey Scilingo pers. comm. 2004
Species at Risk	Henslow's Sparrow	Population fluctuates, 10 ind. in 2004	Breeding	Mickey Scilingo pers. comm. 2004
Responsibility Species Assemblage-Grassland	Killdeer, Upland Sandpiper, Henslow's Sparrow, Bobolink, Eastern Meadowlark	Breed	Breeding	NY BBA 2000
Responsibility Species Assemblage-Shrub/scrub	American Woodcock, Willow Flycatcher, Brown Thrasher, Blue-winged Warbler, Golden-winged Warbler, Eastern Towhee, Field Sparrow	Breed	Breeding	NY BBA 2000

Description:

This site is an active U.S. Army installation. Largely undeveloped, the site holds a variety of habitats including forest, open grasslands, and wetlands. The most prevalent habitats are old fields and young second-growth forest. According to the NY GAP land cover data, approximately 55% of the site is shrub/scrub and open habitats, which include shrub swamp, successional shrub, successional hardwoods, and old field/pasture. The lower elevation southwest corner of the site includes glacially derived features such as recessional moraines, sand plains, drumlins, and dunes, and wetlands. The northeastern third of the site is characterized by a wide zone of foothills and has several lakes, rock outcrops, and many steep-sided, northeast to southwest hillocks.

Blue-winged Warbler

Birds: This site supports significant grassland and shrub breeding bird communities. Breeding species include the Northern Harrier, Upland Sandpiper (30 plus in 1996), American Woodcock, Blue-winged Warbler, Short-eared Owl (periodic nester), Golden-winged Warbler, Prairie Warbler, Vesper Sparrow (100 plus singing males in 2003), Grasshopper Sparrow, and Henslow's Sparrow (12 individuals in 2004). Additional at-risk species breeding on the installation include the American Black Duck, Common Loon, Pied-billed Grebe, American Bittern, Least Bittern (at least one pair), Osprey (at least 1 pair), Sharp-shinned Hawk (few pairs), Cooper's Hawk (at least two pair), Red-shouldered Hawk, Common Nighthawk (several), Whip-poor-will, Red-headed Woodpecker (two pair in 2004), Willow Flycatcher, Horned Lark, Wood Thrush, Cerulean Warbler, and Canada Warbler (possible pair). In 2003 there were at least 35 singing male Clay-colored Sparrows; this species was confirmed breeding in 1998 and its population appears to be increasing.

Conservation: Large portions of this site are kept in an early successional state, to the benefit of the grassland bird community. If the area were abandoned and training maneuvers ceased, the grasslands could be lost to natural succession. Artificial nest programs are ongoing for Wood Ducks, Hooded Mergansers, Tree Swallows, and Eastern Bluebirds. There is concern about the spread of the invasive plant pale swallowwort (*Cynanchum rossicum*). Fort Drum is open to the public for birding and the Natural Resource Branch has created a bird checklist. Contact the Army's Natural Resource Branch for further information on accessing the installation.

Fort Edward Grasslands

Multiple municipalities, Washington County

13,000 acres
150-400' elevation

43.2588°N
73.5423°W

Criteria

Criterion	*Species*	*Data*	*Season*	*Source*
Species at Risk	Northern Harrier	Average 10, max. 17 ind. 1993-2003	Winter	Barb Putnam pers. comm. 2004
Species at Risk	Upland Sandpiper	8 ind. in 1993, 10 in 1994, 6 in 1995, 8 in 1996, 9 in 1998, 5 in 1999, 2 in 2000, 2 in 2001	Breeding	Barb Putnam pers. comm. 2004
Species at Risk	Short-eared Owl	9 ind. in 1993, 14 in 1994, 22 in 1995, 6 in 1996, 8 in 1997, 16 in 1998, 21 in 1999, 12 in 2000, 6 in 2001	Winter	Barb Putnam pers. comm. 2004
Species at Risk	Henslow's Sparrow	4 ind. in 1997, 4 in 1998, 2 in 1999, 7 in 2001	Breeding	Barb Putnam pers. comm. 2004
Responsibility Species Assemblage-Grassland	Killdeer, Upland Sandpiper, Henslow's Sparrow, Bobolink, Eastern Meadowlark	Breed	Breeding	NY BBA 2000

Description: This site consists of a large agricultural grassland complex in the Hudson River Valley. The area contains many working farms and grassland areas interspersed with cultivated fields, small woodlots, and wetlands. According to the NY GAP land cover data, approximately 50% of the site is open habitat, including cropland and old field/pasture land.

Birds: Fort Edward is an exceptional grassland bird breeding and wintering area. The site supports breeding Northern Harriers (two pairs in 2000), Upland Sandpipers (average six, maximum 10 individuals from 1993-1996, 1998-2001), Grasshopper Sparrows (average one, maximum four individuals from 1994-2001), Henslow's Sparrows (four individuals in 1997, four in 1998, two in 1999, and seven in 2001), Vesper Sparrows, Savannah Sparrows (25 plus pairs in 1997), Bobolinks (75 plus pairs in 1997), and Eastern Meadowlarks (10 plus pairs in 1997). It is also an important raptor wintering area with large numbers of Northern

Harriers (average 10, maximum 17 individuals from 1993-2003), Short-eared Owls (average 13, maximum 22 individuals from 1993-2001), and Horned Larks (average 115, maximum 190 individuals from 1993-2003).

Conservation: Much of this area is for sale for residential development. There is an urgent need to find ways to work with agricultural interests and other conservation partners to maintain pastures and other parts of the agricultural landscape. The Audubon Society of the Capitol Region, Southern Adirondack Audubon Society, and the Hudson-Mohawk Bird Club have advocated for the protection of this area by maintaining working farms. Monitoring and inventory of grassland species should continue.

Four Brother Islands

Willsboro, Essex County

15 acres
95' elevation

44.4285°N
73.3348°W

IBA Criteria Met

Criterion	*Species*	*Data*	*Season*	*Source*
Congregations-Waterfowl	Double-crested Cormorant	2,779 estimated pairs in 2003, 2,498 in 2002, 2,437 in 2001, 1,346 in 2000, 1,372 in 1999, 1,394 in 1998, 826 in 1997, 1,184 in 1996, 804 in 1995, 785 in 1994, 532 in 1993	Breeding	John M.C. Peterson pers. comm. 2004
Congregations-Gulls	Ring-billed Gull, Herring Gull, Great Black-backed Gull	10,049 ind. in 2003, 13,058 in 2002, 13,541 in 2001, 14,064 in 2000, 15,085 in 1999, 13,729 in 1998, 15,143 in 1997, 12,610 in 1996, 13,670 in 1995, 15,140 in 1994, 15,131 in 1993	Breeding	John M.C. Peterson pers. comm. 2004
Congregations-Wading Birds	Great Blue Heron, Great Egret, Snowy Egret, Black-crowned Night-Heron, Glossy Ibis	35 estimated pairs in 2003, 76 in 2002, 16 in 2001, 195 in 2000, 277 in 1999, 193 in 1998, 188 in 1997, 132 in 1996, 77 in 1995, 61 in 1994, 13 in 1993	Breeding	John M.C. Peterson pers. comm. 2004

Description: This site includes four islands owned by The Nature Conservancy that are in Lake Champlain approximately 5.5 miles northeast of Willsboro. The four islands, known as A, B, C, and D, consist of shingle-beaches and shale cliffs with vegetated plateaus and clay soils.

Birds: These islands support the largest and most important colonial waterbird nesting colony on Lake Champlain. Maximum numbers include 2,779 pairs of Double-crested Cormorants in 2003, two pairs of Great Blue Herons in 2000, one pair Great Egret of in 2003, four pairs of Snowy Egret in 2003, 20 pairs of Cattle Egrets in 2000, 227 pairs of Black-crowned Night-Herons in 1999, two pairs of Glossy Ibis in 1999, 15,033 individual Ring-billed Gull in 1997, 160 nests of Herring Gulls in 1995, and 10 pairs of Great Black-backed Gulls in 1997.

Conservation: Guano from the colony has killed trees that are used by waterbirds for nesting. In June 1994, 22 waterbirds were illegally shot. Increasing cormorant population may require a management response to protect other nesting species and their habitats. Some birds (adults and chicks) die each year from entanglement in fishing line and lures. The High Peaks Audubon Society has historically hired a warden to conduct a census and bird banding operation; in recent years the University of Vermont performs the surveys. During the first round of IBA site identifications, this site was recognized under the research criterion because a monitoring project is based there.

Great Egret

Indian River/Black Lakes

Multiple municipalities, Jefferson and St. Lawrence Counties

130,000 acres
260-590' elevation

44.4082°N
75.6513°W

IBA Criteria Met

Criterion	*Species*	*Data*	*Season*	*Source*
Species at Risk	Common Loon	3 adults in 2002 [1]; 6 atlas blocks (2 CO, 2 PR, 2 PO) [2]	Breeding	1. Adirondack Cooperative Loon Program; 2. NY BBA 2000
Species at Risk	Osprey	8 atlas blocks (4CO, 1PR, 3PO)	Breeding	NY BBA 2000
Species at Risk	Golden-winged Warbler	10 atlas blocks (4CO, 6PR) [1]; 75 ind. in 1999 [2]	Breeding	1. NY BBA 2000; 2. Matt Young pers. comm 2004
Species at Risk	Cerulean Warbler	8 ind. in 1996, 1 in 1997, 2 in 1998, 11 in 1999, 1 in 2003 [1]; 8 atlas blocks (1CO, 2PR, 5PO) [2]	Breeding	1. Cerulean Warbler Atlas Project; 2. NY BBA 2000
Responsibility Species Assemblage-Grassland	Killdeer, Bobolink, Eastern Meadowlark	Breeds	Breeding	NY BBA 2000
Responsibility Species Assemblage-Shrub/scrub	American Woodcock, Willow Flycatcher, Brown Thrasher, Blue-winged Warbler, Golden-winged Warbler, Eastern Towhee, Field Sparrow	Breeds	Breeding	NY BBA 2000
Congregations-Individual Species	Golden-winged Warbler	Site has supported over 1% of state estimated breeding population; 75 ind. in 1999	Breeding	Matt Young pers. comm. 2004

Description: This site includes the 968-acre Indian River Wildlife Management Area (WMA), administered by NYS DEC, and many privately owned lands. The site extends from Indian River WMA in the south to Black Lake in the north and west to Crooked Creek and Chippewa Creek. The site is a large, diverse area with several lakes and many streams and rivers. The extensive acreage of farmland abandoned 25-30 years ago is now undergoing succession and provides valuable habitat for grassland and shrub birds. According to the NY GAP land cover data, approximately 65% of the site is shrub/scrub and open habitats, which include cropland, old field/pasture, shrub swamp, successional hardwood, and successional shrub land. There are smaller areas of wooded wetlands, grasslands, and active farmland.

Birds: This area contains a variety of wetland as well as agricultural areas, shrub land, and forest. The shrub and grassland habitats support characteristic breeding species. At-risk species that breed here include the Pied-billed Grebe, American Bittern, Least Bittern, Osprey (four pairs), Bald Eagle, Northern Harrier, Whip-poor-will, Sedge Wren (four pairs), Golden-winged Warbler (20-30 pairs), and Cerulean Warbler (four pairs).

Conservation: Golden-winged Warblers have benefited from the large amount of shrub habitat found at this site, which needs to be managed for this species. Development of a plan to work with private landowners to enhance and manage early and mid-successional habitats should be a priority. Potential residential development on higher lands adjoining Chippewa Creek could affect water flow and quality. The Indian River Lakes Conservancy is working on protecting land in this area. Inventory and monitoring, particularly of at-risk species, are needed.

Lake George Peregrine Site

Multiple municipalities, Warren and Washington Counties

475 acres
300-1,620' elevation

43.4691°N
73.6630°W

IBA Criteria Met

Criterion	*Species*	*Data*	*Season*	*Source*
Species at Risk	Peregrine Falcon	5 pairs in 2004, 5 in 2003, 6 in 2002, 5 in 2001, 4 in 2000, 5 in 1999, 4 in 1998, 4 in 1997, 4 in 1996, 3 in 1995, 2 in 1994, 2 in 1993	Breeding	NY Natural Heritage Biodiversity Databases
Congregations-Individual Species	Peregrine Falcon	5 pairs in 2004, 5 in 2003, 6 in 2002, 5 in 2001, 4 in 2000, 5 in 1999, 4 in 1998, 4 in 1997, 4 in 1996, 3 in 1995, 2 in 1994, 2 in 1993	Breeding	NY Natural Heritage Biodiversity Databases

View looking toward Anthony's Nose from Rogers Rock on Lake George

Description: This site includes a series of forested tracts surrounding Lake George, most of which are administered by the NYS DEC or owned by The Nature Conservancy. The numerous outcroppings and cliffs located within this site make it suitable for Peregrine Falcon nesting.

Birds: This area is one of the most unique nesting Peregrine Falcon areas in the state, due to the relatively high number of nesting pairs surrounding a relatively small lake. Over the past ten years, the number of nesting Peregrine Falcons has grown from two to six pairs.

Conservation: Recreational activities can disturb nesting peregrines; NYS DEC posts protection signs at select nesting locations. The Lake George Land Trust is active in acquiring critical habitats, including a recent acquisition near Rogers Rock.

Lisbon Grasslands

Multiple municipalities, St. Lawrence County

60,000 acres
295-395' elevation

44.6607°N
75.2749°W

IBA Criteria Met

Criterion	*Species*	*Data*	*Season*	*Source*
Species at Risk	Black Tern	20+ nests in 2003 [1]; 1 ind. in 1993, 18 in 1994, 34 in 2001 [2]	Breeding	1. Bob Long pers. comm. 2004; 2. Mazzocchi, I.M. and M. Roggie 2004
Responsibility Species Assemblage-Grassland	Killdeer, Bobolink, Eastern Meadowlark	Breed	Breeding	NY BBA 2000

Description: This site includes a vast mosaic of wetlands and grasslands in an agricultural landscape. It contains large areas of wet sedge meadows, which are unusual in most other parts of state. The site includes many privately owned parcels, and the 8,782-acre Upper and Lower Lakes WMA administered by NYS DEC. According to the NY GAP land cover data, approximately 35% of the site is open habitat, including cropland and old field/pasture land.

Birds: As an important wetland complex, this site supports the breeding American Black Duck (confirmed), Common Loon (probable), Pied-billed Grebe (confirmed), American Bittern (probable), Least Bittern (probable), Osprey (confirmed), Northern Harrier (confirmed), American Woodcock (probable), Black Tern (confirmed), Willow Flycatcher (probable), Sedge Wren (confirmed 1990s), Golden-winged Warbler (probable), and Cerulean Warbler (possible). Additional at-rish species include Sharp-shinned Hawk (confirmed), Cooper's Hawk (probable), Common Nighthawk (probable), Horned Lark (probable), Wood Thrush (probable), and Grasshopper Sparrow (probable). The Upper and Lower Lakes WMA has become an important breeding site for Black Terns (20 plus pairs in 2003).

Conservation: The Upper and Lower Lakes WMA is managed for a diversity of wetland-associated species and has been designated as a state Bird Conservation Area. Indian Creek Nature Center is a 300-acre tract of upland and marsh within the WMA that NYS DEC leases to The North Country Conservation Education Association. The center provides a number of programs in environmental education for school groups, scouts, adults, and college classes. The privately owned areas provide good habitat, and landowners should be encouraged to sustain grassland and wetland habitats that will benefit priority species. Inventory and monitoring of at-risk species throughout the area should continue.

Black Tern

Little Galloo Island

Hounsfield, Jefferson County

43 acres
245' elevation

43.8856°N
76.3957°W

IBA Criteria Met

Criterion	*Species*	*Data*	*Season*	*Source*
Congregations-Waterfowl	Double-crested Cormorant	4,251 estimated breeding pairs in 2003, 4,780 in 2002, 5,440 in 2001, 5,119 in 2000, 5,681 in 1999	Breeding	McCullough *et al.* 2004
Congregations-Waterbirds	Ring-billed Gull	60,000 estimated breeding pairs in 2003, 53,000 in 1999	Breeding	McCullough *et al.* 2004
Congregations-Waterbirds	Caspian Tern	1,658 estimated breeding pairs in 2003, 1,585 in 2002, 1,590 in 2001, 1,350 in 2000, 1,440 in 1999	Breeding	McCullough *et al.* 2004
Congregations-Individual Species	Double-crested Cormorant	Largest colony in NY (3,967 in 2004)	Breeding	Irene Mazzocchi pers. comm. 2004
Congregations-Individual Species	Ring-billed Gull	The largest Ring-billed Gull colony in the U.S.	Breeding	McCullough *et al.* 2004
Congregations-Individual Species	Caspian Tern	Only Caspian Tern colony in NY (1,560 pairs in 2004)	Breeding	Irene Mazzocchi pers. comm. 2004

Description: Little Galloo is a small, rocky island located 5.5 miles from Stoney Point and 9.5 miles west of Henderson Harbor, NY, in eastern Lake Ontario. In December 1998, the NYS DEC acquired Little Galloo Island and it became the fourth island within the Lake Ontario Islands WMA. The guano of nesting cormorants has destroyed the few trees on the island; four dead trees (thought to be ash and shagbark hickory) remain, along with a small number of shrubs, and patches of wild geranium, mallow, and other herbaceous vegetation.

Birds: The island hosts an exceptional breeding concentration of colonial waterbirds, including the largest Ring-billed Gull colony in the U.S. (an estimated 60,000 pairs in 2003), New York's only Caspian Tern colony (1,560 pairs in 2004), and the largest Double-crested Cormorant colony in New York (3,967 pairs in 2004). Smaller numbers of Black-crowned Night-Herons (three pairs in 2004), Herring Gulls (313 pairs in 2003), and Great Black-backed Gulls (12 pairs in 2003) nest on the island as well.

Conservation: This site is listed in the 2002 Open Space Conservation Plan as a priority site under the project name Eastern Lake Ontario Shoreline and Islands. A wildlife management plan for the WMA was developed in 2002, with a focus on maintaining the area as a colonial waterbird nesting site. Annual counts and estimates for each species (with the exception of Ring-billed Gulls which, due to large numbers, will be surveyed every five years) will be conducted to understand long term population trends. Productivity estimates will continue annually for all species nesting on the island. Growing cormorant populations have displaced Black-crowned Night-Herons, and management efforts are now in place to reduce the nesting cormorant population to 1,500 pairs. A series of action strategies that include community planning, education, ecosystem management, resource use, and tourism initiatives, are outlined in the plan. Waterfowl hunting has been allowed at the site and will continue. During the first round of IBA site identifications, this site was recognized under the research criterion because a long-term monitoring project is based there.

Lower St. Lawrence River

Multiple municipalities, St. Lawrence County

44,300 acres
165-350' elevation

44.8077°N
75.3269°W

IBA Criteria Met

Criterion	*Species*	*Data*	*Season*	*Source*
Species at Risk	Osprey	Over 5 pairs nesting	Breeding	Lee Harper pers. comm. 2004
Species at Risk	Bald Eagle	64 ind. in winter 2001, 20 in 2000, 28 in 1999, 6 in 1998, 23 in 1997	Winter	NY Natural Heritage Biodiversity Databases
Species at Risk	Common Tern	416 pairs in 2004, 439 in 2003, 418 in 2002, 435 in 2001, 421 in 2000, 435 in 1999, 379 in 1998 [1], 479 in 1997, 455 in 1996, 445 in 1995, 479 in 1994, 502 in 1993 [2]	Breeding	1. Lee Harper pers. comm. 2004; 2. NY Natural Heritage Biodiversity Databases
Congregations-Waterfowl	Mixed species	1,919 ind. ducks, 1,793 Canada Geese in 2002; 4,269 ducks, 12,440 Canada Geese in 2003; 2,180 ducks, 16,778 Canada Geese in 2004	Winter	Ken Ross pers. comm. 2004
Congregations-Waterbirds	Common Tern	79 pairs in 1997, 455 in 1996, 445 in 1995, 479 in 1994, 502 in 1993	Breeding	NY Natural Heritage Biodiversity Databases

Description: Stretching from near Morristown to the Moses Saunders Dam, approximately 60 miles downriver, the Lower St. Lawrence River encompasses a range of wetland and upland habitats. This site is primarily privately owned, but significant acreage is administered by the New York State Power Authority. Additional land is administered by NYS OPRHP and NYS DEC. Two undeveloped islands, Croil Island State Park and Long Sault Island, totaling 1,100 acres, are the largest undeveloped islands in the lower Great Lakes. These islands contain old growth oak forests and significant natural communities, identified by the NY Natural Heritage Program. There are a number of state parks in the area and one state Wildlife Management Area (WMA)—the 3,450-acre Wilson Hill WMA.

Birds: This site supports large numbers of breeding Common Terns (over 450 nests in 12 colonies). There is a very large and globally significant Bank Swallow colony (3,000 plus pairs in 1992; larger than 95% of all Bank Swallow colonies in North America) at Sparrowhawk Point, north of Ogdensburg. Bald Eagles winter along the river (up to 64 counted in 2001), and the area is very important for wintering waterfowl as well. Additional at-risk species supported at the site include the American Black Duck (winter), Common Loon, Pied-billed Grebe, Least Bittern, Northern Harrier, Wood Thrush, and Cerulean Warbler.

Conservation: This site is listed in the 2002 Open Space Conservation Plan as a priority site under the project name St. Lawrence River Islands, Shorelines, and Wetlands. A major factor affecting the health of Common Terns and other fish-eating species is the level of toxins found in the ecosystem. Studies of Common Tern eggs and forage fish in the St. Lawrence from 1986 to 1989 documented the presence of organochlorines, including PCBs, dieldrin, metal and trace elements, including mercury, selenium, copper, aluminum, and cadmium. The levels of several organochlorines and mercury exceeded the no-observed-adverse-effect level and the lowest-observed-adverse-effect level for sensitive species, and may have contributed to low levels of reproductive success. Further work is needed to monitor the reproductive success of other fish-eating birds, and the levels of environmental toxins in the river. The river is a very popular area for recreational boating and fishing, which can potentially disturb birds, especially terns and herons at breeding colonies. Airboats and hovercraft may also disturb wintering birds. Continuing land development pressure is a threat wherever waterfront property exists. Inventory and monitoring, especially of at-risk bird species, should continue.

Massawepie Mire

Colton and Piercefield, St. Lawrence County

5,100 acres
1,480-1,600' elevation

44.2348°N
74.6835°W

IBA Criteria Met

Criterion	*Species*	*Data*	*Season*	*Source*
Species at Risk	Spruce Grouse	12 ind. in summer 2002, 8 ind in. spring 2003	Year-round	Glen Johnson pers. comm. 2004
Responsibility Species Assemblage-Wetland	American Black Duck, Hooded Merganser, American Bittern	Probable breeders	Breeding	Joan Collins pers. comm. 2004
Congregations-Individual Species	Spruce Grouse	Site has supported 10%-40% of NY's Spruce Grouse population	Year-round	Glen Johnson pers. comm. 2004

Spruce Grouse

Description: This large privately-owned boreal peat land in the north central region of the Adirondacks is fringed by spruce and fir forests and bisected by an esker topped with huge white pines. The site is crossed by a dirt road owned by the state, which permits access without harm to the fragile bog habitat. According to the NY GAP land cover data, approximately 40% of the site is wetland habitat, which includes emergent marsh, open fen, wet meadow, evergreen wetland, and shrub swamp.

Birds: This is one of the largest boreal peat lands in the state. Spruce Grouse breed at this site, along with other characteristic northern coniferous forest breeders, including the Black-backed Woodpecker, Yellow-bellied Flycatcher, Gray Jay, Boreal Chickadee, Ruby-crowned Kinglet, Lincoln's Sparrow, Rusty Blackbird, and Pine Siskin. This site has an impressive concentration of breeding Palm Warblers; 30 pairs were observed in 2004 and individuals were estimated to number in the hundreds.

Conservation: This site is listed in the 2002 Open Space Conservation Plan as a priority site under the project name Massawepie Mire. The bog and desirable adjoining properties should be managed in a way that sustains the unique forest habitat. Inventory and monitoring of at-risk breeding birds are needed.

Moose River Plains/Blue Ridge Area

Multiple municipalities, Hamilton and Herkimer Counties

305,000 acres
1,380-3,680' elevation

43.6302°N
74.6016°W

IBA Criteria Met

Criterion	*Species*	*Data*	*Season*	*Source*
Species at Risk	American Black Duck	7CO, 2PR, 5PO [1]; 8CO, 3PR, 8PO [2]	Breeding	1. NY BBA 2000; 2. NY BBA 1980
Species at Risk	Common Loon	25 adults in 2002 [1]; 13CO, 3PR, 4PO	Breeding	1. Adirondack Cooperative Loon Program; 2. NY BBA 2000
Species at Risk	American Bittern	2PR, 2PO [1]; 4PR, 1PO [2]	Breeding	1. NY BBA 2000; 2. NY BBA 1980
Species at Risk	Osprey	4CO, 1PR, 6PO	Breeding	NY BBA 2000
Species at Risk	Sharp-shinned Hawk	Extent of habitat and breeding atlas presence strongly suggest that the threshold is being met [1]; 1PR, 6PO [2]	Breeding	1. Technical Committee consensus; 2. NY BBA 2000
Species at Risk	Cooper's Hawk	Extent of habitat and breeding atlas presence strongly suggest that the threshold is being met [1]; 1PO [2]	Breeding	1. Technical Committee consensus; 2. NY BBA 2000
Species at Risk	Northern Goshawk	Extent of habitat and breeding atlas presence strongly suggest that the threshold is being met [1]; 1CO [2]	Breeding	1. Technical Committee consensus; 2. NY BBA 2000
Species at Risk	Olive-sided Flycatcher	7PR, 3PO [1]; 17PR, 15PO [2]	Breeding	1. NY BBA 2000; 2. NY BBA 1980
Species at Risk	Wood Thrush	Extent of habitat and breeding atlas presence strongly suggest that the threshold is being met [1]; 1PO [2]; 7CO, 12PR, 4PO [3]	Breeding	1. Technical Committee consensus; 2. NY BBA 2000; 3. NY BBA 1980
Species at Risk	Canada Warbler	1CO, 10PR, 4PO [1]; 19CO, 8PR, 11PO [2]	Breeding	1. NY BBA 2000; 2. NY BBA 1980

Criterion	*Species*	*Data*	*Season*	*Source*
Species at Risk	Rusty Blackbird	Extent of habitat and breeding atlas presence strongly suggest that the threshold is being met [1] 10CO, 5PO [2]; 8CO, 2PR, 10PO [3]	Breeding	1. Technical Committee consensus; 2. NY BBA 2000; 3. NY BBA 1980
Responsibility Species Assemblage-Forest	Ruffed Grouse, American Woodcock, Yellow-bellied Sapsucker, Eastern Wood-Pewee, Least Flycatcher, Great Crested Flycatcher, Blue-headed Vireo, Veery, Bicknell's Thrush, Wood Thrush, Northern Parula, Chestnut-sided Warbler, Black-throated Blue Warbler, Black-throated Green Warbler, Blackburnian Warbler, Blackpoll Warbler, Black-and-white Warbler, American Redstart, Ovenbird, Canada Warbler, Scarlet Tanager, Rose-breasted Grosbeak, Purple Finch	Breed	Breeding	NY BBA 2000

Description: This large, relatively intact tract of forest is mostly state-owned and includes the Moose River Plains Wild Forest and Blue Ridge Wilderness Area. The area includes the watershed of the Moose River, a tributary of the Black River. It is generally undisturbed and includes numerous mountains with sub-alpine habitats as well as lakes and ponds. The area holds a diverse group of productive habitats, including dense stands of mixed growth woodlands, flat lands, open woodlands, and black spruce bogs. According to the NY GAP land cover data, approximately 95% of the site is forest, which includes sugar maple mesic, evergreen northern hardwood, deciduous wetland, evergreen plantation, spruce fir, and evergreen wetland forests.

Birds: This area contains some of the best lowland boreal forests and wetlands in the western Adirondacks, and is on the southern periphery of the range of many boreal forest birds. The area supports a number of characteristic boreal birds including the Black-backed Woodpecker, Olive-sided Flycatcher, Yellow-bellied Flycatcher, Gray Jay, Common Raven, and Boreal Chickadee. Within the area there are a number of mountains over 3,500 feet in elevation, and breeding Bicknell's Thrushes have been documented. The area contained the state's last known natural Golden Eagle nesting site.

Conservation: Much of this area has been state designated a Wild Forest and Wilderness Area and is popular for outdoor recreation such as snowmobiling, camping, hiking, fishing, and hunting. Privately-owned portions of this site should be protected from forest-fragmenting development. Sustainable forest management on the private holdings has potential to provide habitat for species requiring successional forest habitats or disturbed forests. Snowmobiling is particularly popular, with 30,000-40,000 visitors participating per year. Illegal ATV use seems to be increasing. More research is needed on whether bird populations are negatively impacted by the current types and levels of recreational use. Acid rain has had a negative impact on the forest and lake ecosystems, though its long-term effects on birds are unclear. Acid rain deposition may be having an impact on the nesting success of songbirds, particularly at high elevations, by killing snails and other edible sources of calcium needed for egg production. More research is needed on this as well. The curtailment of sulphur dioxide emissions and the reduction of acid rain is currently a significant New York State initiative. A detailed inventory and standardized monitoring of at-risk species is needed for the area. Specifically, peaks above 2,800 feet should be surveyed for Bicknell's Thrush.

Olive-sided Flycatcher

Moreau Lake Forest

Multiple municipalities, Saratoga and Warren Counties

10,000 acres
330-1,400' elevation

43.2533°N
73.7464°W

IBA Criteria Met

Criterion	*Species*	*Data*	*Season*	*Source*
Responsibility Assemblage-Forest	Black-billed Cuckoo, Eastern Wood-Pewee, Wood Thrush, Rose-breasted Grosbeak, and Baltimore Oriole	Breed	Breeding	NY BBA 2000
Congregations-Individual Species	Bald Eagle	Site supports 1% (3 ind.) of state wintering eagle population	Winter	Pete Nye pers. comm. 2004

Description: This relatively intact forest, located in Saratoga and Warren counties, is bisected by the Hudson River. This site includes most of the 4,000-acre Moreau Lake State Park, state-designated Forest Preserve lands, and privately owned lands. Moreau Lake State Park contains a seasonal beach and campground, picnic areas, parking areas, and buildings, which make up only 0.75% of the park. The remaining park area is forested upland (93.7%) and other natural areas. According to the NY GAP land cover data, approximately 95% of the site is forested, including sugar maple mesic, evergreen northern hardwood, deciduous wetland, evergreen plantation, oak, successional hardwood, and Appalachian oak pine forest. The site also includes Moreau Lake and about 4.5 miles of the Hudson River.

Birds: The forest habitat supports characteristic birds, including breeding Black-billed Cuckoos, Eastern Wood-Pewees, Wood Thrushes, Rose-breasted Grosbeaks, and Baltimore Orioles. The site is considered an important wintering Bald Eagle site as well.

Conservation: Privately-owned portions of this site should be protected from development that permanently converts forest to non-forest land use and results in forest fragmentation. Sustainable forest management has potential to provide habitat for species requiring successional forest habitats or disturbed forests. A master plan for Moreau Lake State Park, including a recommendation for designating it a Bird Conservation Area, will be completed in 2005.

Northern Adirondack Peaks

Multiple municipalities, Essex and Clinton Counties

29,000 acres
2,800-4,900' elevation

44.3910°N
73.8890°W

IBA Criteria Met

Criterion	*Species*	*Data*	*Season*	*Source*
Species at Risk	Peregrine Falcon	2 pairs in 2002-1999, 1 in 1998-1997, 2 in 1996, 1 in 1995-1993	Breeding	NY Natural Heritage Biodiversity Databases
Species at Risk	Bicknell's Thrush	11 individuals were observed	Breeding	Lambert, J.D. 2003
Congregations-Individual Species	Bicknell's Thrush	11 individuals were observed	Breeding	Lambert, J.D. 2003

Description: This site includes forested areas above 2,800 feet in the northern High Peaks region north of Lake Placid and Keene. The site is almost entirely state-owned and consists of predominately spruce-fir and mixed forest.

Birds: The high elevation habitat found within this site supports breeding Peregrine Falcons and relatively high numbers of Bicknell's Thrushes.

Conservation: Acid rain has had a negative impact on the forest and lake ecosystems, though its long-term effects on birds are unclear. Acid rain deposition may be having an impact on the nesting success of songbirds, particularly at high elevations, by killing snails and other edible sources of calcium needed for egg production. More research is needed on this. The curtailment of sulphur dioxide emissions and the reduction of acid rain is currently a significant New York State initiative. Development projects (e.g., expansion of ski slopes) with potential to disturb Bicknell's Thrush habitat should be monitored closely. A detailed inventory and standardized monitoring of at-risk species is needed for the area. Specifically, peaks above 2,800 feet should be surveyed for Bicknell's Thrush. Some monitoring is conducted by VINS as part of the Mountain Birdwatch program.

Perch River Complex

Multiple municipalities, Jefferson County

66,000 acres
290-500' elevation

44.1046°N
75.9441°W

IBA Criteria Met

Criterion	*Species*	*Data*	*Season*	*Source*
Species at Risk	Bald Eagle	2 pairs breeding	Breeding	NY Natural Heritage Biodiversity Databases
Species at Risk	Northern Harrier	Observed in 8 atlas blocks (1CO, 4PR, 3PO)	Breeding	NY BBA 2000
Species at Risk	Upland Sandpiper	Observed in 5 atlas blocks (3PR, 2PO)	Breeding	NY BBA 2000
Species at Risk	Black Tern	66 ind. in 1994, 118 in 1995, 95 in 1998, 58 in 2001	Breeding	Mazzocchi, I.M. and M. Roggie 2004
Species at Risk	Henslow's Sparrow	14 ind. in 1995, 2 singing males in 1994 [1]; probable breeder in 6 blocks [2]	Breeding	1. NY Natural Heritage Biodiversity Databases; 2. NY BBA 2000
Responsibility Species Assemblage-Shrub	American Woodcock, Willow Flycatcher, Brown Thrasher, Blue-winged Warbler, Eastern Towhee, Field Sparrow,	Breed	Breeding	NY BBA 2000
Responsibility Species Assemblage-Grassland	Killdeer, Upland Sandpiper, Henslow's Sparrow, Bobolink, Eastern Meadowlark	Breed	Breeding	NY BBA 2000
Congregations-Individual Species	Black Tern	Supported over 5% of state breeding population of Black Terns (in 1989, 1990, 1991, 1994, 1998, 2001).	Breeding	Mazzocchi, I.M. and M. Roggie 2004

Description: This site encompasses various wetlands, including flooded valleys, wooded swamps, and wet meadows in the St. Lawrence Valley. Site ownership is a mix of private and state, including NYS DEC administered land (the 7,862-acre Perch River WMA). Three lakes, including Perch Lake, and extensive agricultural grasslands with scattered wet areas adjoin the WMA. According to the NY GAP land cover data, approximately 90% of the site is open and shrub habitat, which includes old field/pasture, shrub swamp, successional hardwood, successional shrub, and cropland.

Birds: This area supports an exceptional wetland bird community, with a diverse array of wetland- and grassland-associated birds. The site has supported one of the largest concentrations of breeding grassland birds in the state. In 1996, there were an estimated one Black Rail, 5-10 Sedge Wrens, 10-20 Grasshopper Sparrows, and 50-70 Henslow's Sparrows within the area. Point count surveys in the area in 1997 tallied 10 plus Upland Sandpipers, 400 plus Savannah Sparrows, 100 plus Grasshopper Sparrows, 80 plus Henslow's Sparrows, 400 plus Bobolinks, and 150 plus Eastern Meadowlarks. Northern Harriers breed here as well (four nests in 2004). Additional species at risk supported at the site include the American Black Duck, Pied-billed Grebe, American Bittern, Least Bittern, Osprey, Bald Eagle, Sharp-shinned Hawk, American Woodcock, Black Tern, Whip-poor-will, Willow Flycatcher, Horned Lark, Sedge Wren, Wood Thrush, and Vesper Sparrow. Many other characteristic wetland species breed here, including Virginia Rails, Sora, Common Moorhens, American Coots, Marsh Wrens, Swamp Sparrows, and many others. Trumpeter Swans have recently bred, possibly originating from birds released in Canada or privately.

Conservation: Portions of this site have been designated as a state Bird Conservation Area. The WMA is managed specifically for wildlife conservation. An inventory of state-listed species was completed in the late 1990s. Continued monitoring of at-risk species is needed. This is one of the state's most important sites for nesting Henslow's Sparrow, and management should take this high priority species into account. A plan is needed to work with farmers to conserve agricultural lands beneficial to grassland birds. Management on state-owned lands should promote grassland and early successional habitat. Invasive plants of concern include pale swallowwort (*Cynanchum rossicum*), purple loosestrife (*Lythrum salicaria*), and common reed (*Phragmites australis*).

Plattsburgh Airfield

Plattsburgh, Clinton County

1,000 acres
125-200' elevation

44.6509°N
73.4657°W

IBA Criteria Met

Criterion	*Species*	*Data*	*Season*	*Source*
Species at Risk	Grasshopper Sparrow	2-17 individuals, average of 9 from 1994-2002; has been declining since 1994	Breeding	Mark Gretch pers. comm. 2003

Grasshopper Sparrow

Description: A former U.S. Air Force Base, this site is now owned by the Plattsburgh International Airport and Clinton County Airport. Areas around the runways are particularly important for birds. The vast majority of the area surrounding the runways consists of dry, barren flats with some grass growing in tufts. There is also some shrub growth scattered throughout the site. Towards the south end of the runway, there are immature pitch pine, small white pines, gray birch, black chokeberry, and some huckleberry bushes.

Birds: Although the number of grassland birds has declined in recent years, this airfield is still an important grassland bird breeding area. Regular bird surveys were carried out at the site each year from 1994-2002. Grasshopper Sparrows declined from 17 in 1994 to two in 2001; Vesper Sparrows from 12 in 1994 to one in 2002; Savannah Sparrows from 22 in 1995 to four in 2000; and Horned Larks from seven in 1996 and 1997 to none seen in 2001. Northern Harriers are regularly sighted foraging at the south end of the runway.

Conservation: Audubon New York provided comments to the current landowners regarding the site's importance to grassland bird populations in 2001 and 2003. Grassland bird species could be extirpated from the site if the area were developed. Rock concert events with tens of thousands in attendance have been held on the airfield. Any direct disturbance to the ground nesting species listed above during the nesting season (May-July) can destroy nests and young. Large numbers of people can cause lasting damage to grassland habitats. For these reasons, the areas that are used by these at-risk species should be off-limits to human intrusion each year until at least August. Continued monitoring of grassland species at the site is needed. Conservation partners are encouraged to work with the airport to ensure that grassland habitat remains viable.

Point Peninsula

Lyme, Jefferson County

6,600 acres
240-295' elevation

43.9997°N
76.2463°W

IBA Criteria Met

Criterion	*Species*	*Data*	*Season*	*Source*
Species at Risk	Northern Harrier	Supports 15-30 ind. or 2-10 ind., depending on vole populations	Winter	Gerry Smith pers. comm. 2004
Species at Risk	Short-eared Owl	Daily max. 30+ ind., varies annually	Winter	Gerry Smith pers. comm. 2004
Species at Risk	Black Tern	20 pairs in 2001, 7 in 1998, 6 in 1994, 30 in 1991, 7 in 1990, 5 in 1989	Breeding	Mazzocchi, I.M. and M. Roggie 2004
Congregations-Waterfowl	Mixed species	Site hosts thousands of waterfowl, with offshore numbers in the tens of thousands	Migration and winter	Gerry Smith pers. comm. 2004
Congregations-Individual Species	Rough-legged Hawk	Supports 1% of estimated state wintering population	Winter	Based on information provided by Gerry Smith
Congregations-Individual Species	Black Tern	Has supported over 1% of state breeding population during 6 years of surveys	Breeding	Mazzocchi, I.M. and M. Roggie 2004
Congregations-Individual Species	Snowy Owl	Supports 1% of estimated state wintering population	Winter	Based on information provided by Gerry Smith
Congregations-Individual Species	Short-eared Owl	Daily max. 30+ ind., varies annually	Winter	Gerry Smith pers. comm. 2004

Description: This site includes a peninsula extending into Chaumont Bay in Lake Ontario's northeast corner, just south of where the St. Lawrence River exits. The site is primarily privately owned, but includes a small NYS DEC Wildlife Management Area and the NYS OPRHP-administered Long Point State Park. The peninsula includes a mix of working and abandoned farms; most of the working farms produce hay. There is a small residential human population with heavier summer usage.

Birds: This site may be one of the most critical winter concentration areas in the northeast for arctic breeding Rough-legged Hawks, Snowy Owls, and Short-eared Owls. During the winter of 1987-88, systematic surveys documented one-day maximums of 57 Northern Harriers, 33 Red-tailed Hawks, 130 Rough-legged Hawks, six American Kestrels, 10 Great Horned Owls, six Snowy Owls, 12 plus Long-eared Owls, 30 plus Short-eared Owls, two Northern Saw-whet Owls, and eight Northern Shrikes. Point Peninsula Shoal, which is offshore, is an important pre-migratory staging area for Caspian Terns, Common Terns, and Black Terns (479 in 1991, 281 in 1994, and 218 in 1996), and also hosts large numbers of waterfowl. Common Terns and Short-eared Owls have bred at the site as well.

Conservation: This site is listed in the 2002 Open Space Conservation Plan as a priority site under the project name Eastern Lake Ontario Shoreline and Islands. Many farms have been abandoned and sold over the past ten years, resulting in a loss of grassland habitats to succession and development. A plan is needed to work with farmers to conserve agricultural lands beneficial to grassland birds.. Additionally, a program should be developed to educate landowners and the public about the importance of grassland habitats to wintering raptors. Disturbance caused by increased recreational ATV use should be investigated. Regular monitoring of wintering raptors is needed. Invasion by swallowwort plants is of great concern; control efforts are underway and need to continue.

Short-eared Owl

Saratoga National Historical Park

Stillwater, Saratoga County

2,885 acres
70-375' elevation

43.0118°N
73.6486°W

IBA Criteria Met

Criterion	*Species*	*Data*	*Season*	*Source*
Species at Risk	Blue-winged Warbler	19 singing males in 2001 [1]; extent of habitat and data source strongly suggests that the threshold is being met [2]	Breeding	1. Jane Graves pers. comm. 2005; 2. Technical Committee consensus
Species at Risk	Henslow's Sparrow	2-15+ individuals observed during breeding season, but not seen since 1999		Linda White pers. comm. 2004
Responsibility Species Assemblage-Grassland	Killdeer, Bobolink, Eastern Meadowlark	Breed	Breeding	NY BBA 2000

Description: Saratoga National Historical Park is located in eastern Saratoga County along the western bank of the Hudson River. The western part of the park consists of low, elongated hills that alternate with broad, flat-bottomed valleys. To the east are two large terraces cut in an east-west direction by deep ravines formed by several creeks. Approximately 27% of the site is grassland. The site played an important role in the American Revolutionary War; two historic battles that stopped a major British advance in 1777 occurred here. The park is a popular recreation and tourism site, with annual visitation of over 350,000 people.

Birds: This site is an important grassland bird breeding area supporting the Northern Harrier, Upland Sandpiper (last observed in 1997 breeding season), Vesper Sparrow (observed in 1997, 2000, 2003, and 2004 breeding seasons), Savannah Sparrow (observed in 2004), Grasshopper Sparrow (last observed in 2001), Henslow's Sparrow (2-15 plus individuals in breeding season, but not seen since 1999), Bobolink, and Eastern Meadowlark. The site also hosts breeding Sharp-shinned Hawks (seen regularly in breeding season and migration), American Woodcock (observed in breeding season in 1997, 2000, 2003-2004), Willow Flycatchers (observed in breeding season in 1997, 2000), Eastern Bluebirds (regularly observed, 60 nest boxes), Blue-winged

Warblers (confirmed breeder), Golden-winged Warblers (observed in breeding season in 1997, 2000, 2003, one individual in 2004), and Prairie Warblers (observed in breeding season). Additional shrub species supported at the site include Eastern Towhee (abundant), Field Sparrow (abundant), and Brown Thrasher.

Conservation: The park has an active research and management program for grassland birds. Grasslands at the site are actively managed to prevent succession. The grasslands contain a large number of grass and plant species, which is a result of past land use, farming, mowing, burning, and soil association and moisture. Grasslands are currently managed through prescribed fire or mowing on a rotational basis. Invasive spotted knapweed plants are replacing native species in some field areas; eradication of such non-native invasives should be a priority. Large special events at the park should not be located in important grassland bird breeding areas during the May through August breeding season. Residential development adjacent to park boundaries is reducing available grassland habitat in the area. Opportunities to protect grasslands surrounding the park should be pursued. Inventory and monitoring of high-priority grassland bird species should continue.

Spring Pond Bog

Multiple municipalities, Franklin and St. Lawrence Counties

4,200 acres
1,480-1,780' elevation

44.3583°N
74.5102°W

IBA Criteria Met

Criterion	*Species*	*Data*	*Season*	*Source*
Species at Risk	Spruce Grouse	2 males in 2002 and 2003	Breeding	Glenn Johnson pers. comm. 2004

An Adirondack bog

Description: This large peat complex includes a 500-acre open bog mat, and a wide diversity of peat land types, including a regionally unique inland raised bog and ladder-form patterned peat land. The acidic, nutrient-poor, waterlogged bog habitat deters the growth of plant species not adapted to such an environment. Characteristic plants include black spruce (*Picea mariana*), tamarack (*Larix laricina*), white-fringed orchid (*Platanthera blephariglottis*), bog rosemary (*Andromeda glaucophylla*), pale laurel (*Kalmia polifolia*), sheep laurel (*Kalmia angustifolia*), and small cranberry (*Vaccinium oxycoccos*).

Birds: This area supports a diverse bird community associated with lowland boreal forest and bog habitats, including the Spruce Grouse, American Three-toed Woodpecker, Black-backed Woodpecker, Olive-sided Flycatcher, Yellow-bellied Flycatcher, Gray Jay, Boreal Chickadee, Palm Warbler, Lincoln's Sparrow, and Rusty Blackbird. In addition, American Black Duck (confirmed breeding in 1997), Common Merganser (confirmed breeding in 1997), and Hooded Merganser have been documented here. Savannah Sparrows were observed in 1996 and 1997, and it was estimated that there were at least three territorial pairs. In total, 130 bird species have been documented here.

Conservation: The site has been owned and managed by The Nature Conservancy since 1985 and is managed specifically to protect its wildlife and plant communities.

Stillwater Reservoir

Webb, Herkimer County

5,800 acres
1,650-1,690' elevation

43.9048°N
74.9754°W

IBA Criteria Met

Criterion	*Species*	*Data*	*Season*	*Source*
Species at Risk	Common Loon	Site supports 5-15 pairs	Breeding	Judy McIntyre's data and Gary Lee pers. comm. 2004
Congregations-Individual Species	Common Loon	Historically, this site has supported over 1% of estimated state breeding population	Breeding	Based on information provided by Judy McIntyre and Gary Lee pers. comm. 2004

Description: Approximately 12 miles north of the Eagle Bay hamlet, this site includes all contiguous bodies of water whose elevation is controlled by the dam system. Most of the site is administered by NYS DEC, but there are some privately owned parcels and other areas under the control of the Hudson River/Black River Water Regulating District. The deciduous and mixed forests surrounding the bodies of water include state designated forest preserve. The dam controls water levels; when the water level is low there are large exposed mudflats. Moose (*Alces alces*) have been observed at this site in recent years.

Birds: This site has had the densest breeding population of Common Loons in the state. In 1995 it supported 15 pairs (5-10% of the state's breeding population). Although the numbers of breeding pairs dropped in 2000 when water levels were lowered, they are now rising again; there were six pairs in 2003. Large numbers of shorebirds have been observed using the exposed muddy lake bottom during spring and fall when water levels are low.

Conservation: The dam retains water for flood control, power generation, navigation, industrial use, and municipal water supply. Water levels clearly determine suitability for nesting loons. The reservoir and surrounding forests are popular recreation areas. Disturbance from recreational boating and jet skis may affect nesting loons. The high mercury levels in reservoir fish may be impacting loons and other fish-eating birds.

Tug Hill Area

Multiple municipalities, Lewis County

79,600 acres
1,280-2,110' elevation

43.6173°N
75.6236°W

IBA Criteria Met

Criterion	*Species*	*Data*	*Season*	*Source*
Responsibility Species Assemblage-Forest	Ruffed Grouse, Black-billed Cuckoo, Yellow-bellied Sapsucker, Eastern Wood-Pewee, Least Flycatcher, Great Crested Flycatcher, Blue-headed Vireo, Veery, Bicknell's Thrush, Wood Thrush, Northern Parula, Chestnut-sided Warbler, Black-throated Blue Warbler, Black-throated Green Warbler, Blackburnian Warbler, Black-and-white Warbler, American Redstart, Ovenbird, Canada Warbler, Scarlet Tanager, Rose-breasted Grosbeak, Purple Finch	Breed	Breeding	NY BBA 2000 and/or Gerry Smith field surveys

Description: Tug Hill is a relatively unfragmented landscape located between the eastern end of Lake Ontario and the Adirondack Mountains. It is ecologically distinct from the Adirondacks and Catskills. Alkaline shale and sandstone-based soils help buffer the area from acid atmospheric deposition. The land of this gently sloping plateau is owned by a variety of private individuals, non-profit organizations, corporations, and local and state governments. Tug Hill is the source of 11 different rivers and streams, and is a mosaic of diverse and extensive wetlands. A remote core area of wetlands and spruce-northern hardwood forest uninterrupted by paved roads is found at the highest elevations (1,700-2,100'). According to the NY GAP land cover data, approximately 90% of the site is forested, which includes deciduous wetland, evergreen northern hardwood, evergreen plantation, evergreen wetland, sugar maple mesic, and successional hardwood forests

Birds: This site supports a number of characteristic forest breeders, including the Ruffed Grouse, Black-billed Cuckoo, Yellow-bellied Sapsucker, Eastern Wood-Pewee, Least Flycatcher, Great Crested Flycatcher, Blue-headed Vireo, Veery, Bicknell's Thrush, Wood Thrush, Northern Parula, Chestnut-sided Warbler, Black-throated Blue Warbler, Black-throated Green Warbler, Blackburnian Warbler, Black-and-white Warbler, American Redstart, Ovenbird, Canada

Warbler, Scarlet Tanager, Rose-breasted Grosbeak, and Purple Finch. At-risk species that have been documented at the site in recent years include the American Black Duck, American Bittern, Sharp-shinned Hawk, Northern Goshawk, American Woodcock, Wood Thrush, and Canada Warbler. Bald Eagles have historically bred at this site.

Conservation: A portion of this site is listed in the 2002 Open Space Conservation Plan under the project name Tug Hill Core Forests and Headwater Streams. In 2002, The Nature Conservancy (TNC) purchased approximately 44,300 acres in the East Branch of the Fish Creek watershed. TNC retained ownership of approximately 14,000 acres of this total, and sold approximately 30,300 acres to a private timber company, GMO Renewable Resources. TNC retained a conservation easement on the 30,300 acres, and title to 1,350 acres of riparian buffer along the East Branch of the Fish Creek to help protect this area. TNC plans to transfer the 30,300-acre conservation easement and 1,350-acre riparian buffer to NYS DEC, and convey a conservation easement to NYS DEC that will permanently protect the 14,000 acres they currently own. The Tug Hill Tomorrow Land Trust holds easements on approximately 1,000 acres. Threats to this area include unsustainable logging, residential and camp development, and ATV use in sensitive areas. Additional protection and outreach to private landowners is needed to encourage sustainable forestry and responsible recreational use.

Black-and-white Warbler

Upper St. Lawrence/Thousand Islands

Multiple municipalities, Jefferson and St. Lawrence Counties

68,000 acres
235-370' elevation

44.2864°N
76.0727°W

IBA Criteria Met

Criterion	*Species*	*Data*	*Season*	*Source*
Species at Risk	Common Loon	1CO, 1PR, 2PO; 2CO, 1PR, 1PO	Breeding	NY BBA 2000; NY BBA 1980
Species at Risk	Osprey	5+ pairs breeding	Breeding	Lee Harper pers. comm. 2004
Species at Risk	Common Tern	315 pairs in 2004, 263 in 2003, 273 in 2002, 202 in 2001, 204 in 2000, 138 in 1999, 168 in 1998 [1]; 163 in 1997; 208 in 1996 [2]	Breeding	1. Lee Harper pers. comm. 2004; 2. NY Natural Heritage Biodiversity Databases
Responsibility Species Assemblage-Shrub	American Woodcock, Willow Flycatcher, Brown Thrasher, Eastern Towhee, Field Sparrow	Breed	Breeding	NY BBA 2000
Congregations-Waterfowl	Mixed species	7,532 ind. in 1994 [1]; 1,826 ind. Common Mergansers in 1996 [2]	Winter	1. Lee Harper pers. comm. 2004; 2. NYS DEC mid-winter aerial waterfowl surveys
Congregations-Waterbirds	Common Tern	Site regularly supports over 100 ind [1]; 163 pairs in 1997, 208 pairs in 1996 [2]	Breeding	1. Lee Harper pers. comm. 2004; 2. NY Natural Heritage Biodiversity Databases

Description: This site stretches from Wilson Bay on Cape Vincent where Lake Ontario drains into the St. Lawrence, 42 miles downriver to Chippewa Bay. It encompasses a range of wetland, shrub, and upland habitats. According to the NY GAP land cover data, 20% of the site is shrub habitat, which includes old field/pasture, successional shrub, successional shrub, and small amounts of shrub swamp. There are numerous islands, ranging in size from large islands like Wellesley and

Grindstone Islands to the many small islands in Chippewa Bay. The site is primarily privately owned, but some parcels are administered by NYS DEC, NYS OPRHP, and owned by the Thousand Island Land Trust (TILT).

Birds: This site is an important Common Tern nesting area, supporting 315 pairs in 2004. It is also an important waterfowl migration and wintering area, with thousands of individuals documented. Large numbers of Common Goldeneyes and Common Mergansers use this stretch of the river (1,826 Common Mergansers were counted in an aerial survey in January 1996) and, in the past, this area has had large concentrations of Canvasbacks and Redheads. The site supported the largest Great Blue Heron rookery in the state at Ironsides Island (1,147 nests in 1995), but there were no nests in 2004. There are probably still hundreds of Great Blue Herons in the area in scattered rookeries. At-risk breeding species include the American Black Duck (winters and breeds), Common Loon (breeds), Least Bittern (possible breeder), Osprey (breeds), Bald Eagle (winters), Northern Harrier (breeds), Sharp-shinned Hawk (possible breeder), Red-shouldered Hawk (possible breeder), American Woodcock (probable breeder), Common Tern (breeds), Black Tern (historic breeder), Whip-poor-will (probable breeder), Red-headed Woodpecker (possible breeder), Willow Flycatcher (possible breeder), Wood Thrush (probable breeder), Cerulean Warbler (confirmed breeder, two nests observed on TILT property), and Vesper Sparrow (possible breeder). This is an important Bald Eagle wintering area, with 8-12 individuals seen each winter, especially near Wellesley Island.

Conservation: This site is listed in the 2002 Open Space Conservation Plan as a priority site under the project name St. Lawrence River Islands, Shorelines, and Wetlands. A major factor affecting Common Terns and other fish-eating species is the level of toxins found in the ecosystem. Studies of Common Tern eggs and forage fish in the St. Lawrence from 1986-1989 documented the presence of organochlorines, including PCBs, dieldrin, metals, and trace elements, including mercury, selenium, copper, aluminum, and cadmium. The levels of several organochlorines and mercury exceeded the no-observed-adverse-effect level and the lowest-observed-adverse-effect level for sensitive species and may contribute to low levels of reproductive success. Further work is needed to monitor the reproductive success of other fish-eating birds and the levels of environmental toxins in the river. The river is a very popular area for recreational boating and fishing, which can disturb birds, especially terns and herons at breeding colonies. Development of shoreline

View of the St. Lawrence River

habitats is also a concern. Save the River, a non-profit, member-based environmental organization whose mission is to preserve and protect the ecological integrity of the Thousand Islands Region of the St. Lawrence River, recently adopted this IBA. Save the River has an active program that monitors and protects Common Tern nesting sites. The Thousand Island Land Trust is also active in this area, purchasing grasslands, riparian areas, and wet woods; they currently own about 3,300 acres of land and hold conservation easements on 3,500 acres.

Valcour Island

Plattsburgh and Peru, Clinton County

1,100 acres
90-190' elevation

44.6216°N
73.4166°W

IBA Criteria Met

Criterion	*Species*	*Data*	*Season*	*Source*
Congregations-Wading Birds	Great Blue Heron	416 active nests in 2004, 369 in 2003, 421 in 2002, and 552 in 2001	Breeding	Richards and Capen 2004

© Jeff Nadler

Great Blue Herons on nest

Description: This 1,100-acre island is in the Towns of Peru and Plattsburgh in Lake Champlain. The island is owned by the state and administered by NYS DEC. It is located within Adirondack Park and classified as primitive, which means it is managed to achieve and maintain a condition as close to wilderness as possible. There are a number of rare plant species. The island is accessible by boat and is an important anchorage site for mariners on Lake Champlain. It is used for camping (there are a limited number of camp sites around the perimeter of the island), hiking, wildlife viewing, and hunting. There is also a historic lighthouse located on Bluff Point, on the western shore of the island.

Birds: This site supports the largest Great Blue Heron rookery on Lake Champlain; it is also the third largest rookery in the Great Lakes region (2004). It has been active since at least the 1980s. Surveys performed by the University of Vermont (UVM) documented 552 active nests in 2001, 421 in 2002, 369 in 2003, and 416 in 2004. Since the UVM surveys are performed after the breeding season ends, the count numbers are considered a minimum, as nests may have fallen down due to weather and wind.

Conservation: This site is not expected to undergo significant increases in human use or development. There is some concern that high levels of recreational use might disturb the nesting herons, although the current location of the rookery makes it difficult to access. There is also concern that Double-crested Cormorants may start nesting on the island; cormorant guano can destroy trees used by nesting Great Blue Herons.

William C. Whitney Wilderness Area

Long Lake, Hamilton County

21,000 acres
1,715-2,300' elevation

44.0066°N
74.6722°W

IBA Criteria Met

Criterion	*Species*	*Data*	*Season*	*Source*
Species at Risk	Common Loon	9 adults in 2002	Breeding	Adirondack Cooperative Loon Program
Responsibility Species Assemblage-Wetland	American Black Duck, Hooded Merganser, American Bittern	Breed	Breeding	NY BBA 2000

Description: This site is located in the ecological transition zone between the temperate deciduous forest and the boreal forest. The main forest types are northern hardwoods (mostly beech, red maple, and yellow birch), mixed woods (hardwoods with hemlock, red spruce, and scattered white pine), and spruce flats. The site contains 11 lakes, including Little Tupper Lake, which is nearly six miles long and up to a mile wide. The landscape surrounding the lakes is composed of low, forested hills with a few modest mountains. Elevations above sea level range from 1,717 feet at the surface of Little Tupper Lake to 2,297 feet at the summit of Antediluvian Mountain. Extensive wetlands stretch out from the ponds and streams. According to the NY GAP land cover data, approximately 30% of the site is wetland habitat, which includes emergent marsh/open fen/wet meadow, evergreen wetland, and shrub swamp. Except for three private holdings, the site is state-owned.

Birds: This site provides exceptional wetland habitat for characteristic species. Probable and possible at-risk species include the American Black Duck, American Bittern, Bald Eagle, Northern Goshawk, Canada Warbler, and Rusty Blackbird. Confirmed at-risk breeders include the Pied-billed Grebe, Osprey, and Olive-sided Flycatcher. Palm Warblers also breed here.

Conservation: A NYS DEC ranger is assigned to this site to help protect and manage it as a wilderness area. A unit management plan will be developed. The site provides vast areas for recreation. Heavy logging activity prior to state ownership has left a young, open forest over much of the area. A narrow band of mature trees was preserved along the shorelines to protect the view from the water and to reduce erosion into ponds and streams. Acid rain has had a negative impact on the forest and lake ecosystems, though its long-term effects on birds are unclear. Acid rain deposition may be having an impact on the nesting success of songbirds, particularly at high elevations, by killing snails and other edible sources of calcium needed for egg production. More research is needed on this. The curtailment of sulphur dioxide emissions and the reduction of acid rain is currently a significant New York State initiative. A detailed inventory and standardized monitoring of at-risk species are needed for the area.

© Jeff Nadler

American Bittern

Chapter 3.3 Catskills and Hudson River Valley

Hudson Highlands

Catskills and Hudson River Valley

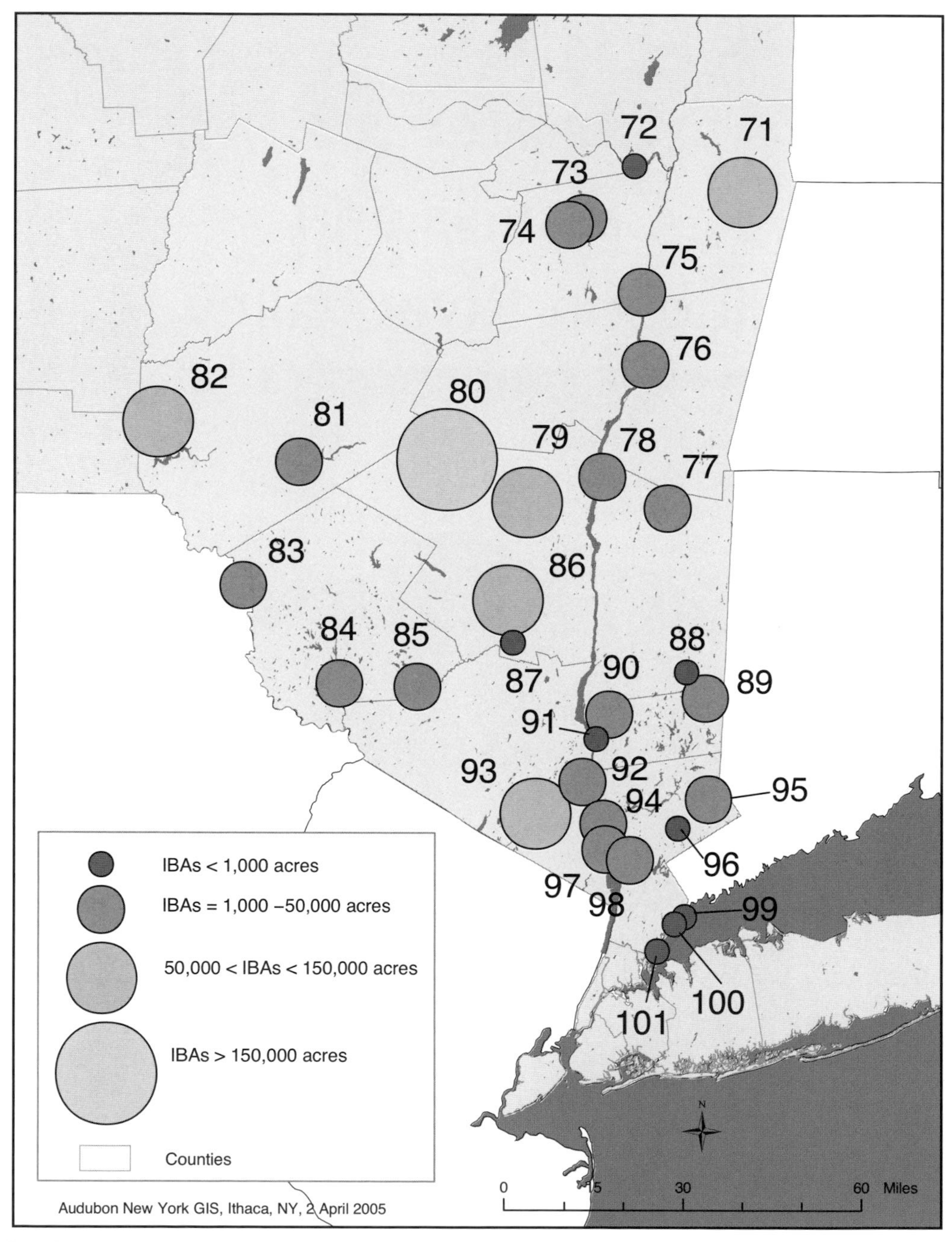

Fig. 3.3.1 IBAs in the Catskills and Hudson River Valley

Catskills and Hudson River Valley

Ashokan Reservoir Area

Multiple municipalities, Ulster County

70,000 acres
200-2,440' elevation

41.9864°N
74.1754°W

IBA Criteria Met

Criterion	*Species*	*Data*	*Season*	*Source*
Species at Risk	American Black Duck	325 ind.	Winter	Maynard Vance, NYS DEC
Species at Risk	Common Loon	50-75 ind.	Spring and fall migrations	Frank Murphy pers. comm. 2004
Species at Risk	Red-shouldered Hawk	Estimate 20 pairs	Breeding	Frank Murphy pers. comm. 2004
Responsibility Species Assemblage-Forest	Sharp-shinned Hawk, Black-billed Cuckoo, Eastern Wood-Pewee, Wood Thrush, Rose-breasted Grosbeak, Baltimore Oriole	Breeds	Breeding	NY BBA 2000

Description: This site is located in the northern part of Ulster County and includes the Ashokan reservoir and extensive surrounding forest habitat. The reservoir is approximately 9,000 acres and is part of the New York City water supply system, owned by the New York City Department of Environmental Protection. Surrounding lands are generally undisturbed beech/oak/maple and mixed pine/hemlock forest, much of which is protected as a buffer zone for the reservoir. According to the NY GAP land cover data, approximately 85% of the site is forested, and includes Appalachian oak-pine, deciduous wetland, evergreen northern hardwood, evergreen plantation, oak, and sugar maple mesic forests. The area has been largely undisturbed since 1915.

Birds: This site has supported a nesting pair of Bald Eagles for the past ten years and hosts up to six Bald Eagles during the winter. The reservoir also serves as a stopover site for waterfowl, including an estimated 325 American Black Ducks and 50 or more Common Loons. Surrounding woodlands support an estimated 20 pairs of breeding Red-shouldered Hawks, as well as other breeding at-risk birds, including American Black Ducks, American Woodcocks, Whip-poor-wills, Red-headed Woodpeckers, Willow Flycatchers, Wood Thrushes, Blue-winged Warblers, Prairie Warblers, and Worm-eating Warblers. Migrating at-risk species include Pied-billed Grebes (potential breeder), Ospreys, Sharp-shinned Hawks, Cooper's Hawks, and Peregrine Falcons. In the fall, when the reservoir is low, the site also supports shorebirds, including American Golden-Plovers, Pectoral Sandpipers, White-rumped Sandpipers, and Baird's Sandpipers. Flocks of Snow Buntings and Lapland Longspurs also use the site during the fall.

Conservation: This site is listed in the 2002 Open Space Conservation Plan as a priority site under the project name New York City Watershed Lands. The surrounding area is under heavy development pressure, due in large part to its close proximity to New York City. The reservoir is a major water supply for New York City and has been protected since 1915. It is open to fishing and there are hiking and hunting access points surrounding the reservoir. Occasionally, logging has taken place on the city-owned parcels. New York City's 2001 *Watershed Protection Program Summary, Assessment, and Long-term Plan* has been published. Perhaps the most direct human impact is on nesting eagles by anglers. Signs forbidding entry near eagle nests have been posted. Monitoring of disturbance to eagles by anglers should continue.

Bashakill Wildlife Management Area

Deerpark and Mamakating, Orange and Sullivan Counties

2,360 acres
485-655' elevation

41.5364°N
74.5192°W

IBA Criteria Met

Criterion	*Species*	*Data*	*Season*	*Source*
Congregations-Waterfowl	Mixed species, but especially important for Canada Geese and Wood Ducks	Regularly over 2,000 ind.	Migration	Sullivan County Audubon Society

Description: This site is administered by the New York State Department of Environmental Conservation (NYS DEC) and includes the extensive emergent wetlands that surround the Bashakill River. The primary habitat is non-tidal wetlands surrounded by deciduous woods and mixed woods. The southeast side is bordered by an abandoned railroad right-of-way and the Delaware and Hudson canal borders the northeast side. Some boardwalks and limestone caves exist in the area. The site hosts a number of rare plants, including the spreading globeflower (*Trollius laxus*), and rare animals, including the ironcolor shiner (*Notropis chalybaeus*), known at only 10 sites in the state, and long-tailed salamander (*Eurycea longicauda*). The area also contains one of the largest bat hibernacula in the state, with six species of bats.

Birds: The Bashakill is an important wetland complex, especially for the downstate area, hosting fall concentrations of Canada Geese (5,000) and Wood Ducks (1,000-2,000). It also hosts many at-risk species, including American Black Ducks (migrant), Pied-billed Grebes (breed), American and Least Bitterns (migrants), Ospreys (breed), Bald Eagles (1 pair), Northern Harriers (migrant), Sharp-shinned Hawks (breed), Cooper's Hawks (breed), Northern Goshawks (breed), Red-shouldered Hawks (breed), and Rusty Blackbirds (migrant). Characteristic wetland-dependent species like Great Blue Herons, Virginia Rails, Sora, and Common Moorhens breed here as well.

Wood Ducks

Conservation: This site is listed in the 2002 Open Space Conservation Plan as a priority site under the project name Neversink Highlands and Shawangunk Mountains. This site has been designated a state Bird Conservation Area. NYS DEC staff released insect control agents in 2001 and 2002 to manage the invasion of purple loosestrife (*Lythrum salicaria*). Due to limited staff time, there has not been active monitoring to measure the success of these efforts. Purple loosestrife is still present in the wetland. Local birders have expressed concern about declining numbers of some waterfowl species as a result of altered water levels and increased vegetation growth. Inventory and monitoring, especially of waterfowl and at-risk species, are needed. There are also concerns about developments within the watershed. The Sullivan County Audubon Society hosts regular educational field trips and organizes cleanup events.

Black Creek Marsh

Guilderland and New Scotland, Albany County

1,000 acres
350-390' elevation

42.6712°N
73.9659°W

IBA Criteria Met

Criterion	Species	Data	Season	Source
Species at Risk	American Bittern	Supports 5-10 pairs (based on calling males)	Breeding	Larry Alden pers. comm. 2004
Species at Risk	Least Bittern	Extent of habitat and other data strongly suggest that the threshold is being met [1]; 1 ind. heard 2000, 1998, 1995 [2]; 1 adult and 1 immature July 2003 [3]	Breeding	1. Technical Committee consensus; 2. NY Natural Heritage Biodiversity Databases; 3. Jane Graves pers. comm. 2005

Description: This large wetland complex associated with the Black Creek and its tributaries is surrounded by upland forest and agricultural fields. Nearly half of the site (450 acres) is a Wildlife Management Area managed by the NYS DEC; the rest is privately owned. The old Delaware & Hudson Railroad line divides the marsh into northern and southern sections.

Birds: This site supports at-risk species including the American Black Duck (possible breeder), Pied-billed Grebe (migrant), American Bittern (breeds), Least Bittern (breeds), Northern Harrier (can been seen year round, possible breeder), American Woodcock (breeds), Short-eared Owl (rare winter visitor), Common Nighthawk (possible migrant), Willow Flycatcher (probable breeder, potentially supports 10 pairs), Blue-winged Warbler (probable breeder in low numbers), Prairie Warbler (probable breeder in low numbers), and Rusty Blackbird (migrant, 40-50 ind.). Additional wetland associated species include the Great Blue Heron, Green Heron, Virginia Rail, Sora, Common Moorhen, Wilson's Snipe, Alder Flycatcher (probable breeder), and Marsh Wren. A great diversity of species occurs in the mix of habitats here, especially during the spring migration, when 90-100 species can be identified in one day. Concentrations of several hundred waterfowl can occur in spring and fall, and Canada Geese, Wood Ducks, Mallards, Ruffed Grouse, Wild Turkeys, and American Woodcocks are common breeders.

American Kestrels, Brown Thrashers, Eastern Bluebirds, and Bobolinks occur at the site and/or in adjoining fields. Reports from 1960s indicate that the Henslow's Sparrow, Grasshopper Sparrow, and Sedge Wren occurred here.

Conservation: This site is listed in the 2002 Open Space Conservation Plan as a priority site under the project name Black Creek Marsh/Vly Swamp. The NYS DEC manages 450 acres for wildlife conservation, of which a 65-acre field is being managed for grassland species. About 70 acres are owned by the Town of Guilderland, and several hundred acres of the Indian Ladder Farm (which adjoins the Black Creek Marsh to the south) has recently been put under a conservation easement and will remain as a working farm and orchard. However, loss of open space and uplands adjoining the marsh are continually occurring and can be expected to intensify. Much of the upland and farmland adjoining the marsh is for sale and likely to be developed. Current local land use and zoning is not adequate to protect the important wildlife habitats at the site. Conversion of adjoining lands to housing would be detrimental to the marsh habitat and wetland bird populations. Exploration of potential ways to protect adjoining lands should be initiated as a partnership between local groups and state agencies. Better monitoring of wetland bird species at the site is needed.

Butler Sanctuary

Bedford, New Castle, North Castle, Westchester County

356 acres
360-790' elevation

41.1816°N
73.6857°W

IBA Criteria Met

Criterion	*Species*	*Data*	*Season*	*Source*
Congregations-Raptors	Mixed species	11,508 ind. in 1993, 13,355 in 1994, 17,798 in 1995, 12,105 in 1996, 8,351 in 1997, 9,075 in 1998, 14,419 in 1999, 17,428 in 2000, 9,631 in 2001	Fall migration	Hawk Migration Association of North America, Drew Panko pers. comm. 2003

Description: A ridge and swale area in a heavily developed area of Westchester County. The two ridges that compose the bulk of the sanctuary run in a north-south direction and are 700 or more feet in elevation. The hawk watch takes place on an observation platform located at the highest point (above 775') on the easternmost ridge. From the hawk watch, Long Island Sound can be seen. This site is owned by The Nature Conservancy.

Birds: The hawk watch counted a fall average of 12,630 hawks from 1993-2001. The highest season total was 31,077 in 1986. Relatively high numbers of Ospreys (642 in 1987), Sharp-shinned Hawks (4,942 in 1986), Broad-winged Hawks (23,069 in 1986), American Kestrels (1,329 in 1989), and Merlins (86 in 1996) have been counted.

Conservation: Bedford Audubon Society has sponsored a hawk watch, which was not conducted in 2003 and 2004, but is expected to resume in 2005. Monitoring of hawks should continue. Bird monitoring in the surrounding forests is recommended as well.

Sharp-shinned Hawk

Cannonsville/Steam Mill Area

Multiple municipalities, Broome, Chenango, and Delaware Counties

65,000 acres
985-2,075' elevation

42.1629°N
75.3502°W

IBA Criteria Met

Criterion	*Species*	*Data*	*Season*	*Source*
Species at Risk	Bald Eagle	4 pairs in 2002, 3 in 2001, 3 in 2000, 3 in 1999, 3 in 1998, 3 in 1997, 2 in 1996, 2 in 1995, 2 in 1994, 2 in 1993	Breeding	NY Natural Heritage Biodiversity Databases
Responsibility Species Assemblage-Forest	Northern Flicker, Eastern Wood-Pewee, Least Flycatcher, Wood Thrush, Black-throated Blue Warbler, Black-and-white Warbler, Canada Warbler, Scarlet Tanager	Breed	Breeding	NY BBA 2000 and Audubon NY statewide IBA field surveys

Bald Eagle

Description: This site includes the Cannonsville Reservoir, which lies in the former Delaware River valley, an additional mile of river to the north and two miles to the south, and the surrounding contiguous forest. According to the NY GAP land cover data, approximately 70% of the site is forested, and includes Appalachian oak-pine, deciduous wetland, evergreen northern hardwood, evergreen plantation, oak, sugar maple mesic, and successional hardwood forests. The reservoir and surrounding land are owned by the City of New York and are part of the public water supply.

Birds: Bald Eagles are found at the site throughout the year and have nested here since 1988. Other at-risk species supported at the site include the American Black Duck (migrant), Common Loon (migrant), Osprey (migrant), Wood Thrush (breeds), Prairie Warbler (breeds), and Canada Warbler (breeds). The reservoir serves as a stopover site for waterfowl including the Tundra Swan, Wood Duck, American Wigeon, Blue-winged Teal, Northern Pintail, Ring-necked Duck, Black Scoter, Bufflehead, Common Goldeneye, Common Merganser, Red-breasted Merganser, and Horned Grebe.

Conservation: This site is listed in the 2002 Open Space Conservation Plan as a priority site under the project name New York City Watershed Lands. Permanent protection and stewardship of private portions of the site are needed to prevent fragmentation resulting from development. Options include public or land trust acquisition, purchase of conservation easements, and sustainable forestry agreements. The major human impact on eagles at the site is caused by recreational boaters and anglers. Signs forbidding entry near eagle nests have been posted. Monitoring of disturbance levels to eagles should continue.

Catskills Peaks Area

Multiple municipalities,
Ulster, Greene, Sullivan, and Delaware Counties

310,000 acres
600-3,950' elevation

42.0716°N
74.4639°W

IBA Criteria Met

Criterion	*Species*	*Data*	*Season*	*Source*
Species at Risk	Cooper's Hawk	Extent of habitat and BBA data strongly suggest that the threshold is being met [1]; 5PO [2]; 3PO [3]	Breeding	1. Technical Committee consensus; 2. NY BBA 2000; 3. NY BBA 1980
Species at Risk	Northern Goshawk	Extent of habitat and BBA data strongly suggest that the threshold is being met [1]; 1CO, 1PO [2]; 5PO [3]	Breeding	1. Technical Committee consensus; 2. NY BBA 2000; 3. NY BBA 1980
Species at Risk	American Woodcock	Extent of habitat and BBA data strongly suggest that the threshold is being met [1]; 2PR, 4PO [2]; 3CO, 1PR, 1PO [3]	Breeding	1. Technical Committee consensus; 2. NY BBA 2000; 3. NY BBA 1980
Species at Risk	Bicknell's Thrush	21 ind. recorded in 2002	Breeding	Lambert, J.D. 2003
Species at Risk	Wood Thrush	3CO, 10PR, 9PO [1]; 8CO, 10PR, 15PO [2]	Breeding	1. NY BBA 2000; 2. NY BBA 1980
Species at Risk	Canada Warbler	Extent of habitat and BBA data strongly suggest that the threshold is being met [1]; 2CO, 1PR, 4PO [2]; 5CO, 12PR, 10PO [3]		1. Technical Committee consensus; 2. NY BBA 2000; 3. NY BBA 1980

Criterion	*Species*	*Data*	*Season*	*Source*
Responsibility Species Assemblage-Forest	Sharp-shinned Hawk, Black-billed Cuckoo, Northern Flicker, Eastern Wood-Pewee, Least Flycatcher, Yellow-throated Vireo, Blue-gray Gnatcatcher, Wood Thrush, Black-throated Blue Warbler, Cerulean Warbler, Black-and-white Warbler, Louisiana Waterthrush, Canada Warbler, Scarlet Tanager, Rose-breasted Grosbeak	Breed	Breeding	NY BBA 2000
Congregations-Individual Species	Bicknell's Thrush	Site supports over 1% of state breeding population	Breeding	Lambert, J.D. 2003

Description: This site is located within the Catskill Park and includes one of the largest contiguous forest tracts in the state. The coniferous forests above 3,500 feet are primarily composed of balsam fir (*Abies balsamea*) and red spruce (*Picea rubens*). The lower elevation hardwood forests are dominated by sugar maple (*Acer saccharum*) and beech (*Fagus grandifolia*). According to the NY GAP land cover data, approximately 97% of the site is forested, and includes Appalachian oak-pine, deciduous wetland, evergreen northern hardwood, evergreen plantation, evergreen wetland, spruce-fir, oak, and sugar maple mesic forests. Most of the site is administered by NYS DEC, but some is privately owned as well.

The Catskills

Birds: The Catskill peaks over 3,000 feet support a distinctive sub-alpine bird community including breeding Yellow-bellied Flycatchers, Swainson's Thrushes, Hermit Thrushes, Magnolia Warblers, Yellow-rumped Warblers, White-throated Sparrows, and Dark-eyed Juncos. Peaks over 3,500 feet support breeding Bicknell's Thrushes and Blackpoll Warblers. This is the southernmost extension of the breeding range of these two species. In 1997, spot-mapping surveys of three plots totaling 24.4 hectares on two Catskill peaks (two on Hunter and one on Plateau) found 11 Yellow-bellied Flycatcher territories, 23-27 Bicknell's Thrush territories, 10-11 Swainson's Thrush territories, 12-13 Magnolia Warbler territories, 22-25 Yellow-rumped Warbler territories, and 71-78 Blackpoll Warbler territories (Rimmer and McFarland 1997). The average density of Bicknell's Thrush over the three study plots in 1997 was 50 territories/40 hectares. Researchers estimated that Hunter Mountain might have supported 30-35 pairs of Bicknell's Thrush in 1997. Other at-risk species found at the site include the American Black Duck (breeds), Osprey (breeds), Bald Eagle (observed in breeding season), Sharp-shinned Hawk (breeds), Cooper's Hawk (breeds), Northern Goshawk (breeds), Red-shouldered Hawk (breeds), Peregrine Falcon (breeds), American Woodcock (breeds), Olive-sided Flycatcher (breeds), Wood Thrush (breeds), Cerulean Warbler (breeds), and Canada Warbler (breeds).

Conservation: This site is listed in the 2002 Open Space Conservation Plan as a priority site under the project name Catskill Unfragmented Forest. Portions of this site have been designated a state Bird Conservation Area. High elevation habitats are relatively secure throughout the Catskills because they are state-owned and protected under the state constitution. However, a proposal was put forth in 1996 to amend the constitution to allow the Hunter Mountain Ski Bowl to lease and develop 500 acres of high elevation habitat. The proposal was tabled, but the ski area owners continue to argue that the development is necessary for the business to remain economically viable. Permanent protection of the remaining privately owned portions is needed to prevent their development and conversion to non-forest uses. Options include public acquisition and conservation easements. Working forest easements that provide for sustainable forestry may provide suitable habitats for species that require successional or disturbed forests. Recently there has been concern over a proposed development on Belleayre Mountain that, if completed, would fragment this intact forest. The Catskill Peaks are very popular for recreation—particularly hiking and, in some localities, mountain biking. More research is needed on whether bird populations are significantly impacted by current levels of human visitation. More research is also needed on possible impacts that acid rain may be having on the nesting success of songbirds, particularly at high elevations. Acid rain can kill snails, whose shells provide a source of calcium needed for egg production in birds; it can also kill trees, thereby degrading songbird habitat. The curtailment of sulfur dioxide emissions and the reduction of acid rain is currently a significant NY State initiative. A detailed inventory and standardized monitoring of at-risk species is needed for the area. In particular, all peaks above 3,500 feet should be surveyed for Bicknell's Thrush.

Constitution Marsh Sanctuary

Phillipstown, Putnam County

350 acres
0-100' elevation

41.4067°N
73.9444°W

IBA Criteria Met

Criterion	*Species*	*Data*	*Season*	*Source*
Species at Risk	American Black Duck	400-700 ind. in migration	Migration	Eric Lind pers. comm. 2004
Species at Risk	Least Bittern	2-4 breeding pairs each year	Breeding	Eric Lind pers. comm. 2004

Description: This site consists of a 4,000-5,000 year old fresh and brackish (depending on the time of year) tidal marsh (270 acres) and forested uplands (80 acres) located on the east shore of the Hudson River, directly opposite West Point Military Academy, and 52 miles north of New York City. There are a series of human-made dikes and channels that were constructed in the 1830s for wild rice farming within the marsh. The site is administered by the New York State Office of Parks, Recreation and Historic Preservation (NYS OPRHP) and managed by Audubon New York.

Birds: This important wetland site hosts a diversity of birds (more than 200 species have been identified). Characteristic wetland breeders at the site include Least Bitterns (2-4 pairs each year), Virginia Rails, Marsh Wrens, and Swamp Sparrows. Large numbers of waterfowl use the area during winter and migration, with average fall concentrations of 1,500 individuals and occasional peak counts of 2,000 that can include 700 Wood Ducks and several hundred American Black Ducks and Mallards. Mixed flocks of blackbirds (Bobolinks, Red-winged Blackbirds, and Common Grackles) numbering in the thousands, use the site as a staging area and migratory stopover in the fall. Other at-risk species using the site include Pied-billed Grebes (occasional migrants), American Bitterns (uncommon but regular migrants), Ospreys (regular migrants, non-breeding visitors), Bald Eagles (averaging 2-5 in winter, with a maximum of 30), Northern Harriers (regular migrants), Sharp-shinned Hawks (fairly common foragers), Cooper's Hawks (probable breeders), Red-shouldered Hawks (rare migrants), Merlins (regular migrants), Peregrine Falcons (occasional foragers), Willow Flycatchers (estimated 3-5 breeding pairs), Wood

Thrushes (breed in adjacent woodlands), Blue-winged Warblers (possible breeders), Cerulean Warblers (regular migrants), Worm-eating Warblers (breed in adjacent woodlands), and Canada Warblers (regular migrants). Until mid-1990s, fall swallow concentrations at the site typically numbered about 20,000 individuals, but reached as high as 100,000. Today, swallow concentrations number in the thousands.

Conservation: This site is listed in the 2002 Open Space Conservation Plan as a priority site under the project name Hudson River Corridor Estuary/ Greenway Trail. The site is part of the Hudson Highlands State Park and is managed by Audubon New York as a wildlife conservation area. Portions have been designated as a state Bird Conservation Area. There is an Audubon Center on site that provides education programs to thousands of people each year. Non-native invasive plants and animals that require monitoring include common reed *(Phragmites australis)*, purple loosestrife *(Lythrum salicaria)*, European water chestnut *(Trapa natans)*, zebra mussels *(Dreissena polymorpha)*, and Mute Swans. The sanctuary is part of a federal Superfund Site and cadmium and nickel contamination have been remediated. Regular monitoring of contaminant levels is ongoing.

© Eric Lind

Canoeing at Constitution Marsh

Doodletown and Iona Island

Multiple municipalities, Orange and Rockland Counties

1,675 acres
0-1,085' elevation

41.2980°N
73.9900°W

IBA Criteria Met

Criterion	*Species*	*Data*	*Season*	*Source*
Species at Risk	Cerulean Warbler	7-8 ind. in 1997, 11 in 1998	Breeding	Cerulean Warbler Atlas Project

Description: Situated along the western shore of the Hudson River just south of West Point, this site includes portions of Bear Mountain State Park, and encompasses deciduous forest and riverine habitats and freshwater/ brackish tidal wetlands. From the cattail marshes along the river, the land slopes steeply upward, with occasional small streams lined with hemlocks. Oaks, and cottonwoods, with an understory of barberry, dominate most of the forest. The Doodletown portion of this site is an abandoned settlement with scrubby, secondary growth. The two areas support populations of rare wildlife like the timber rattlesnake (*Crotalus horridus*), and rare dragonflies including Needham's skimmer (*Libellula needhami*), arrowhead spiketail (*Cardulegaster obligua*), comet darner (*Anex longipes*), and gray petaltail (*Tachopteryx thoreyi*). Unusual plants include sedges such as *Carex buschii, Carex emonsii,* and *Carex seorsa,* along with yellow corydalis (*Corydalis flavula*), flat sedge (*Cyperus odoratus*), and frost grape (*Vitis vulpina*).

Birds: At-risk species supported at this site include the Pied-billed Grebe (migrant), Osprey (migrant), Bald Eagle (use Iona Island in winter), Northern Harrier (migrant), and Golden-winged Warbler (possible breeder, seen occasionally near reservoir in Doodletown). Other species documented here include the Acadian Flycatcher, Louisiana Waterthrush, Kentucky Warbler, Hooded Warbler, and many other common species. Doodletown supports an unusual diversity and abundance of breeding warblers and other songbirds. More than 165 bird species have been documented at the site. Iona Island provides wetland habitat for characteristic species.

Iona Island

Conservation: This site is listed in the 2002 Open Space Conservation Plan as a priority site under the project name Highlands Greenway Corridor. Doodletown and Iona Island have been designated as a state Bird Conservation Area. In the early 1990s, common reed (*Phragmites australis*) took over most of Iona marsh, displacing much of the cattail. Iona marsh can be viewed from an unimproved parking lot off of the causeway; other access is limited to guided canoe and kayak trips. Any improvements or expansions of park facilities on Iona Island need thorough assessments of impacts on birds, especially Bald Eagles.

Edith G. Read Wildlife Sanctuary

Rye, Westchester County

270 acres
0-50' elevation

40.9691°N
73.6702°W

IBA Criteria Met

Criterion	Species	Data	Season	Source
Species at Risk	American Black Duck	30-40 pairs nesting	Breeding	Jason Klein pers. comm. 2004
Congregations-Waterfowl	Mixed species	Site regularly supports 3,500-4,500 waterfowl during winter, including as many as 3,000 Greater Scaup [1]; supports large numbers of scaup, 5,500 ind. at one time in 2004, max. of 12,000 seen in early 1990s [2]	Winter	1. Usai, M.L 1996; 2. Jason Klein pers. comm. 2004

Description: This site is a wildlife sanctuary located in a heavily urbanized area along the north shore of Long Island Sound. The central habitat feature is the 80-acre Playland Lake, a brackish lake located within Rye Playland amusement park. The area also includes a one-quarter mile stretch of shore with a rocky intertidal community on Long Island Sound, two miles of walking trails through various successional states of maritime forest woodlands, wetland and open field habitats, and informal gardens. The site is owned by the Westchester County Department of Parks, Recreation, and Conservation.

Birds: The site hosts a remarkable diversity of migrant and wintering birds. The lake is an important wintering area for waterfowl, harboring thousands of birds in mixed species flocks. The area is particularly important as a wintering area for Greater Scaup, supporting 8-10% of the state's wintering population of the species (5,500 in December 2004). Other waterfowl species observed include White-winged Scoters, Surf Scoters, and Northern Gannets (100 individuals offshore in 2004). The site is used as a foraging area for healthy numbers of wading birds (50-100 individuals at a time), including Great Blue Herons, Great Egrets, Snowy Egrets, and Black-crowned Night-Herons (8 pairs). More than 290 species have been recorded here, and the site is a migrant songbird stopover with records of many unusual species. At-risk species supported here include Common Loons (six in December

1994), Pied-billed Grebes (two in November 1996), Northern Harriers (one in May 1995), Red-shouldered Hawks (one in November 1994), American Woodcocks (at least three pairs), Common Terns (12 in July 1995), Black Terns (three in May 1995), Black Skimmers (two in June 1995), Willow Flycatchers (breed), and Yellow-breasted Chats (breed; one in December 1995).

Conservation: This site is listed in the 2002 Open Space Conservation Plan as a priority site under the project name Westchester Marine Corridor. Overuse by patrons of Rye Playland is a concern, and steps have been taken to develop a more cooperative approach between the sanctuary and the amusement park to increase sensitivity to the natural importance of the sanctuary. A project led by the Army Corps of Engineers to reintroduce tidal flow to the lake is being considered. Recent habitat enhancements include dune restoration, spartina re-introduction, invasive species management, and meadow restoration. Monitoring of waterfowl and at-risk species should continue.

Fahnestock and Hudson Highlands State Parks

Multiple municipalities, Dutchess and Putnam Counties

14,800 acres
0-1,400' elevation

41.4476°N
73.8516°W

IBA Criteria Met

Criterion	*Species*	*Data*	*Season*	*Source*
Responsibility Species Assemblage-Forest	Northern Flicker, Eastern Wood-Pewee, Least Flycatcher, Yellow-throated Vireo, Blue-gray Gnatcatcher, Wood Thrush, Black-and-white Warbler, Worm-eating Warbler, Louisiana Waterthrush, Hooded Warbler, Scarlet Tanager, Rose-breasted Grosbeak	Breed	Breeding	New York State Office of Parks, Recreation and Historic Preservation 1995

Description: This rugged area (elevation change from sea level, along the Hudson River, to over 1,400 feet on the summit of Mt. Taurus) lies in a heavily wooded section of Putnam County, a very developed part of the state, and provides a resource for hiking, skiing, swimming, boating, fishing, bow hunting, camping, and other recreational activities. It includes the largest state park in the Taconic region, Fahnestock State Park, and the Hudson Highlands State Park. The Taconic Outdoor Education Center educates school and recreational groups and presents a variety of public programs. Within Fahnestock State Park, the Hubbard/Perkins Conservation Area is a large unfragmented forest tract. The area includes six lakes, a hemlock stream ravine, and marsh habitat. Much of the forest is oak and mixed hardwoods, with an understory of mountain laurel. According to the NY GAP land cover data, approximately 90% of the site is forested, and includes Appalachian oak-pine, deciduous wetland, evergreen northern hardwood, evergreen plantation, oak, and sugar maple mesic forests. Also present are relatively large stands of hemlock and white pine.

Birds: The deciduous and mixed forests support a representative bird community. Breeding birds include Ruffed Grouse, Sharp-shinned Hawks (confirmed breeders), Cooper's Hawks (confirmed breeders), Red-shouldered Hawks (at least three breeding sites), Broad-winged Hawks, Northern Goshawks (foragers, probable breeders), Barred Owls, Whip-poor-wills (at least four breeding sites), Acadian Flycatchers, Yellow-throated Vireos, Blue-headed Vireos, Warbling Vireos, Winter Wrens, Blue-gray Gnatcatchers, Veeries, Hermit Thrushes, Blue-winged Warblers, Black-throated Green Warblers, Blackburnian Warblers (at least two locations), Prairie Warblers, Cerulean Warblers, Worm-eating Warblers, Kentucky Warblers (at least two locations), Canada Warblers (at least one location), Ovenbirds, Northern and Louisiana Waterthrushes, Hooded Warblers, Scarlet Tanagers, and Dark-eyed Juncos. The granite cliffs provide nesting sites for Peregrine Falcons and Common Ravens. The adjacent Hudson River supports migrating shorebirds, ducks, geese, and a variety of other waterbirds.

Conservation: This site is listed in the 2002 Open Space Conservation Plan as a priority site under the project name Fahnestock State Park and the Highlands Greenway Corridor. Portions of the Fahnestock State Park have been designated a state Bird Conservation Area. There are some potential pollution problems from oil spills and dumping in the Hudson River. Hiking traffic can be very heavy during the warmer months and could pose a threat to nesting Peregrine Falcons, so should be carefully monitored. Some wetlands have populations of invasive common reed (*Phragmites australis*), which should be monitored. Increased mountain bike traffic and the illegal use of all-terrain vehicles may be problematic and need to be monitored. Inventory and monitoring of breeding birds, especially at-risk species, are needed.

Great Swamp

Multiple municipalities, Dutchess and Putnam Counties

4,550 acres
400-690' elevation

41.5401°N
73.5980°W

IBA Criteria Met

Criterion	*Species*	*Data*	*Season*	*Source*
Congregations-Shorebirds	Mixed species	Solitary Sandpipers migrate along the swamp, probably numbering about 400-500 per day during the peak of migration	Spring migration	Jim Utter pers. comm. 2004

Description: The Great Swamp stretches for approximately 20 miles between the parallel ridges of the Taconic Range. The dominant plant community is red maple swamp, although the immediate area contains floodplain forest and savannah, shrub swamp, shallow marsh, wet meadow, sedge and shrub fen, as well as streams and ponds. The site ranks as the second largest freshwater wetland in the state. It supports several plant communities and species that are unusual in the state, including one of few populations of spreading globeflower (*Trollius laxus)* and perhaps the largest population of lizard's tail (*Saururus cernuus*). The wetlands also support one of the last breeding sites for bog turtles in Putnam and Dutchess counties, and healthy populations of blue-spotted salamanders, wood turtles, painted turtles, and river otters.

Birds: This site contains a large, high quality wetland that supports an exceptional representative bird community. The site is also important for migrating shorebirds, including the Greater Yellowlegs, Solitary Sandpiper, Spotted Sandpiper, Wilson's Snipe, and American Woodcock. At-risk species supported at the site include the American Black Duck (regular migrant), Common Loon (migrant), Pied-billed Grebe (fairly common during migration and winter if the water do not freeze), American Bittern (present in two locations during the breeding season), Least Bittern (has bred, but known nesting sites have been flooded by beaver activity), Osprey (migrant), Bald Eagle (seen in winter), Northern Harrier (migrant and winter), Sharp-shinned Hawk (common during migration and winter), Cooper's Hawk (estimated four pairs), Northern Goshawk (breeds in uplands), Red-shouldered Hawk (breeder and migrant), American Woodcock (regular breeder and migrant), Common Nighthawk (migrant, 10-50 individuals),

Whip-poor-will (breeds in surrounding areas), Willow Flycatcher (migrant and regular breeder), Horned Lark (migrant and winter), Wood Thrush (migrant and regular breeder), Blue-winged Warbler (migrant and regular breeder), Prairie Warbler (migrant and regular breeder in watershed), Bay-breasted Warbler (migrant), Cerulean Warbler (migrant and breeder), Worm-eating Warbler (migrant and breeder), Canada Warbler (migrant and small number breed), Yellow-breasted Chat (migrant, possible breeder), Vesper Sparrow (migrant), and Rusty Blackbird (regularly flocks during migration, occasionally seen in winter).

Conservation: This site is listed in the 2002 Open Space Conservation Plan as a priority site under the project name Great Swamp. The Great Swamp is located along a major development corridor along Route 22 and the northern extension of I-684 and as such is a prime target for road widening and development. Rail service to New York City also runs through the swamp and has been upgraded with new stations and expanded parking lots. Residential and commercial developments continue to be initiated throughout the watershed, including areas immediately adjacent to the swamp. The Regional Plan Association produced a conservation plan for the south flow (East Branch Croton River) portion of the Great Swamp in 1991. This raised public awareness about the value of the swamp, led to the formation of the Friends of the Great Swamp (FrOGS), and served as an impetus for other activities in the area. Both counties have designated the site a Critical Environmental Area, which affords the site a higher level of protection. In 1999, The Nature Conservancy completed a conservation management plan under a U.S. Environmental Protection Agency grant. Since the south flow is the headwaters of the Croton part of the New York City Reservoir System, the New York City Department of Environmental Protection has some planning and regulatory authority over this portion of the site. Very little of the land is in public ownership, but small parcels, including the Town of Patterson Environmental Park, Green Chimneys School, and a parcel owned by Putnam County, allow canoe access to the river. The NYS DEC annually stocks trout since there is no sustainable breeding population. Local outdoors clubs have established a Wood Duck nest box program along the river that appears to have been successful. In 2002, a coalition of partners, led by FrOGS, received a $900,000 grant from the North American Wetlands Conservation Council to assist in the protection of land within the Great Swamp. A complete inventory of birds at the site is needed, as is monitoring of at-risk species.

Harriman and Sterling Forests

Multiple municipalities, Orange and Rockland Counties

63,800 acres
295-1,340' elevation

41.2298°N
74.1386°W

IBA Criteria Met

Criterion	*Species*	*Data*	*Season*	*Source*
Species at Risk	Cooper's Hawk	Extent of habitat and BBA data strongly suggest that the threshold is being met [1]; 5CO [2]	Breeding	1. Technical Committee consensus; 2. NY BBA 2000
Species at Risk	Wood Thrush	Extent of habitat and BBA data strongly suggest that the threshold is being met [1]; 3CO, 3PR [2]; 6CO, 1PO [3]	Breeding	1. Technical Committee consensus; 2. NY BBA 2000; 3. NY BBA 1980
Species at Risk	Blue-winged Warbler	Extent of habitat and BBA data strongly suggest that the threshold is being met [1]; 3CO, 3PR [2]; 4CO, 2PR, 1PO [3]	Breeding	1. Technical Committee consensus; 2. NY BBA 2000; 3. NY BBA 1980
Species at Risk	Golden-winged Warbler	Habitat models estimate 149 pairs [1]; 22 pairs 1997 [2]	Breeding	1. John Confer pers. comm. 2004; 2. Nomination, John C. Yrizarry
Responsibility Species Assemblage-Forest	Sharp-shinned Hawk, Northern Flicker, Eastern Wood-Pewee, Least Flycatcher, Yellow-throated Vireo, Blue-gray Gnatcatcher, Wood Thrush, Black-throated Blue Warbler, Cerulean Warbler, Black-and-white Warbler, Worm-eating Warbler, Louisiana Waterthrush, Hooded Warbler, Scarlet Tanager, Rose-breasted Grosbeak	Breed	Breeding	NY BBA 2000

Harriman State Park

Description: This large contiguous forest is part of the Hudson Highlands and has a strong relief ranging from around 300 to over 1,300 feet in elevation. Due to its geographic location, climax forest communities are oak/maple/beech forests with hemlocks present in higher elevation ravines. According to the NY GAP land cover data, approximately 90% of the site is forested, and includes Appalachian oak-pine, deciduous wetland, evergreen northern hardwood, oak, and sugar maple mesic forests. NYS OPRHP administers Harriman State Park (over 46,000 acres) and Sterling Forest State Park (over 17,000 acres), but this IBA also includes privately owned lands.

Birds: This site supports a healthy representative community of forest breeders, including the Sharp-shinned Hawk, Cooper's Hawk, Northern Goshawk, Red-shouldered Hawk, Broad-winged Hawk, Northern Flicker, Eastern Wood-Pewee, Acadian Flycatcher, Least Flycatcher, Yellow-throated Vireo, Brown Creeper, Winter Wren, Blue-gray Gnatcatcher, Hermit Thrush, Wood Thrush, Black-throated Blue Warbler, Cerulean Warbler, Black-and-white Warbler, Worm-eating Warbler, Ovenbird, Louisiana Waterthrush, Hooded Warbler, Scarlet Tanager, Rose-breasted Grosbeak, and Purple Finch. Additional at-risk species supported at this site include the Osprey (possible breeder), Bald Eagle (winters, eight individuals in 2003 and three in 2002), American Woodcock (probable breeder), Whip-poor-will (breeder), Olive-sided Flycatcher (possible breeder), Blue-winged Warbler (confirmed breeder), Golden-winged Warbler (confirmed breeder), and Prairie Warbler (confirmed breeder).

Conservation: A portion of this site is listed in the 2002 Open Space Conservation Plan as a priority site under the project name Sterling Forest. Portions of Sterling Forest have been designated a state Bird Conservation Area. Residential development on privately owned lands is a concern due to potential habitat fragmentation and loss of habitat for species at risk. Permanent protection and stewardship of private portions of the site are needed to prevent fragmentation resulting from development. Options include public or land trust acquisition, purchase of conservation easements, and sustainable forestry agreements. Regular inventory and monitoring, particularly of at-risk species, should continue. The site is one of the few where Blue-winged Warblers and Golden-winged Warblers occur together in an apparently stable ratio. In other portions of their range, Golden-winged Warblers are undergoing rapid declines with replacement by Blue-winged Warblers. In most areas where these two species occur, the habitat becomes a sink for Golden-winged Warblers. Research into how the two species are coexisting at Sterling Forest is ongoing and could be critical to preventing the loss of Golden-winged Warblers as a breeding species in the state and the region. Habitat restoration for Golden-winged Warblers in the Indian Hill area of Sterling Forest State Park is planned for 2005. Over-browsing by deer and invasion by non-native vegetation have significantly altered the forest in much of this area.

Hook Mountain

Clarkstown and Haverstraw, Rockland County

1,800 acres
60-670' elevation

41.1440°N
73.9177°W

IBA Criteria Met

Criterion	Species	Data	Season	Source
Congregations-Raptors	Mixed species	Fall season totals: 6,551 ind. in 1993, 9,966 in 1994, 13,095 in 1995, 6,054 in 1996, 9,822 in 1997, 13,644 in 1998, 21,234 in 1999, 10,447 in 2000, 19,161 in 2001	Fall migration	Hawk Migration Association of North America, Drew Panko pers. comm. 2003

Description: This site includes almost seven continuous miles of riverfront and cliff slopes in Upper Nyack. Hook Mountain is the site of a fall hawk watch that has operated annually at the precipice of a ridge since 1971. The site is administered by NYS OPRHP, with some small, privately owned parcels adjoining the park.

Birds: Averages of more than 12,000 hawks (maximum 25,929) are counted here each fall. This includes averages of 269 Ospreys (maximum 435), 11 Bald Eagles (maximum 48), 183 Northern Harriers (maximum 362), 3,380 Sharp-shinned Hawks (maximum 7,070), 85 Cooper's Hawks (maximum 166), 20 Northern Goshawks (maximum 112), 97 Red-shouldered Hawks (maximum 184), 3 Golden Eagles (maximum seven), and 13 Peregrine Falcons (maximum 54). Other hawk species include averages of 7,603 Broad-winged Hawks (maximum 18,733), 232 Red-tailed Hawks (maximum 462), 540 American Kestrels (maximum 930), and 29 Merlins (maximum 59).

Conservation: A privately owned, 12-acre meadow abuts a 16-acre wooded wetland at the base of the south face of Hook Mountain. Both properties are known resting and feeding areas for migratory songbirds and hawks, and both are vulnerable to development. Monitoring of hawk flights at the site should continue. During the first round of IBA site identifications, this site was recognized under the research criterion because it is a site where a long-term monitoring project is based.

Huckleberry Island

New Rochelle, Westchester County

15 acres
10' elevation

40.8886°N
73.7571°W

IBA Criteria Met

Criterion	*Species*	*Data*	*Season*	*Source*
Congregations-Wading Birds	Mixed species	28 nests in 2004, 118 in 2003, 99 in 2002, 140 in 2001, 80 in 2000 [1]; 87 pairs in 1998, 103 in 1995, 426 in 1993 [2]	Breeding	1. NY City Audubon Harbor Herons report; 2. Long Island Colonial Waterbird and Piping Plover Survey
Congregations-Individual Species	Double-crested Cormorant	Has supported 25% in 1998, 20% in 1995, 20% in 1993 of the Long Island breeding population	Breeding	NY City Audubon Harbor Herons report and Long Island Colonial Waterbird and Piping Plover Survey
Congregations-Individual Species	Great Egret	Has supported 10% in 1998, 7% in 1995, 10% in 1993 of the Long Island breeding population	Breeding	NY City Audubon Harbor Herons report and Long Island Colonial Waterbird and Piping Plover Survey
Congregations-Individual Species	Black-crowned Night-Heron	Has supported 3% in 1998, 3% in 1995, 10% in 1993 of the Long Island breeding population	Breeding	NY City Audubon Harbor Herons report and Long Island Colonial Waterbird and Piping Plover Survey

Description: This site is a small, privately owned island with a rocky shore, mature deciduous forest, and brush habitat (on the northern peninsula). There are a few small, unused buildings on the southwest end of the island.

Birds: The island supports one of the few waterbird colonies in Long Island Sound. Surveys in 1998 documented that the island supported 25% of the state's coastal breeding Double-crested Cormorants (660 pairs), 10% of the state's Great Egrets (58 pairs), and 3% of the state's Black-crowned Night-Herons (27 pairs). Historically, the island has supported larger numbers of colonial waterbirds, including 114 pairs of Great Egrets in 1989, 145 pairs of Snowy Egrets in 1988, 290 pairs of Black-crowned Night-Herons in 1989, 2,405 pairs of Herring Gulls in 1988, and 546 pairs of Great Black-backed Gulls in 1988. Numbers of Black-crowned Night-Heron (11 nests in 2004) and Snowy Egret (5 nests in 2004) declined further in 2004. Small numbers of Green Herons, Yellow-crowned Night-Herons, Glossy Ibis, and American Oystercatchers have bred here as well.

Conservation: This site is listed in the 2002 Open Space Conservation Plan as a priority site under the project name Westchester Marine Corridor. Cormorant droppings have been killing many trees at this site, making them unsuitable for nesting by herons and egrets. There is some concern that the non-native Norway maple (*Acer platanoides*) understory may replace the native hardwoods. Also, there is potential for the site to be used for recreational purposes that would disrupt nesting birds. New York City Audubon has surveyed this site as part of their Harbor Herons surveys and monitoring should continue. Obtaining permanent protection of this site is strongly encouraged.

John Boyd Thacher State Park

Multiple municipalities, Albany County

1,270 acres
590-1,540' elevation

42.6477°N
74.0105°W

IBA Criteria Met

Criterion	*Species*	*Data*	*Season*	*Source*
Species at Risk	Wood Thrush	Extent of habitat strongly suggests that the threshold is being met	Breeding	Technical Committee consensus
Species at Risk	Canada Warbler	Extent of habitat and other data strongly suggest that the threshold is being met [1]; max of 8 singing males on a walk [2]	Breeding	1. Technical Committee consensus; 2. Jane Graves pers. comm. 2004

John Boyd Thacher State Park

Description: This site includes a large area of northern hardwood and mixed forest along and atop the Helderberg Escarpment; it is administered by NYS OPRHP. Jefferson and spotted salamanders migrate in spring from below the top of the escarpment to the ponds in the center of the park. The escarpment contains world-renowned fossil beds and provides spectacular views of the valley.

Birds: The forest habitat supports some of the Albany area's highest densities of breeding songbirds, including the Winter Wren, Hermit Thrush, Wood Thrush, Magnolia Warbler, Black-throated Blue Warbler, Black-throated Green Warbler, Blackburnian Warbler, Worm-eating Warbler, Louisiana and Northern Waterthrush, and Canada Warbler. The cliffs historically harbored breeding Peregrine Falcons and could still support this species. The site supported the first Common Raven nest in the region, and is now the nucleus for the area's population. A large roost of Turkey Vultures can be seen in the spring and fall. At-risk species supported at the site include the Bald Eagle (rare), Northern Harrier (breeding evidence), Sharp-shinned Hawk (breeds), Cooper's Hawk (breeds, year-round resident), Northern Goshawk (breeds, year-round resident), Red-shouldered Hawk (uncommon), Peregrine Falcon, American Woodcock (breeds), Common Nighthawk (uncommon migrant), Olive-sided Flycatcher (uncommon migrant), Willow Flycatcher (breeds), Wood Thrush (breeds), Blue-winged Warbler (breeds, common migrant), Golden-winged Warbler (breeds), Prairie Warbler (breeds), Bay-breasted Warbler (uncommon migrant), Worm-eating Warbler (breeds), Canada Warbler (breeds), and Vesper Sparrow (rare).

Conservation: This site is listed in the 2002 Open Space Conservation Plan as a priority site under the project name Helderberg Escarpment. Portions of this site have been designated as a state Bird Conservation Area. Residential and other development in areas immediately adjoining the park may increase fragmentation and predation by human-associated predators, including cats and raccoons.

Little Whaley Lake

Pawling, Dutchess County

800 acres
685-1,085' elevation

41.5640°N
73.6490°W

IBA Criteria Met

Criterion	Species	Data	Season	Source
Species at Risk	Golden-winged Warbler	4+ ind.	Breeding	Nomination, Sibyll Gilbert

Description: Formerly the site of a Boy Scout camp, this privately owned site includes a 45-acre unspoiled natural lake surrounded by a nearly mature mixed deciduous-hemlock forest, with an abundant mountain laurel understory. The site also holds numerous small vernal pools, wooded wetlands, steep slopes, and limey bedrock outcroppings; it has great botanical diversity.

Birds: An exceptional representation of deciduous and mixed hemlock forest with the associated breeding bird communities. Breeding species (with estimates of average numbers) include the Northern Goshawk (two in breeding season), Cooper's Hawk (two in breeding season), Osprey (one to two in migration), Swainson's Thrush, Hermit Thrush, Golden-winged Warbler (four plus in breeding season), Cerulean Warbler (two in breeding season), Blackburnian Warbler, and Canada Warbler.

Conservation: This site is listed in the 2002 Open Space Conservation Plan as a priority site under the project name Northern Putnam Greenway. The area is owned by a developer, though no development or land sales have taken place. The site has been designated a Significant Natural Area by Dutchess County and a Critical Environmental Area by the Town of Pawling. Preservation and drafting of a sensible management plan for the site should be encouraged. More inventories and monitoring of at-risk species are needed to better understand bird use and determine if this site continues to meet IBA criteria.

Lower Hudson River

Multiple municipalities, Duchess, Orange, Putnam, Rockland, and Westchester Counties

28,000 acres
0-1,500' elevation

41.2881°N
73.9627°W

IBA Criteria Met

Criterion	*Species*	*Data*	*Season*	*Source*
Species at Risk	Bald Eagle	Winter ind.: 8 in 2003, 3 in 2002, 22 in 2001, 35 in 2000, 14+ in 1998, 6 in 1997, 60 in 1996, present in 1995; Breeding pairs: 2 in 2002, 2 in 2001, 1 in 2000	Breeding and winter	NY Natural Heritage Biodiversity Databases
Congregations-Individual Species	Bald Eagle	Supports about 10% of state winter Bald Eagle population	Winter	NY Natural Heritage Biodiversity Databases

Description: This site includes the Lower Hudson River, extending just north of the Newburgh-Beacon Bridge south to, and including, Croton Point Park. Some additional winter roost sites for eagles in the surrounding uplands are also included. This site includes state, county, and private ownership.

Birds: One of the most critical wintering Bald Eagle sites in the state, and becoming an important breeding area for Bald Eagles. Croton Point Park supports wintering and breeding grassland birds including the Northern Harrier (year-round), Short-eared Owl (up to six individuals in winter), Grasshopper Sparrow (observed in breeding season), Vesper Sparrow (observed during migration), and Henslow's Sparrow (observed during migration).

Conservation: This site is included in the 2002 Open Space Conservation Plan under the project name Hudson River Corridor Estuary and Greenway Trail. Inventory and monitoring of eagles should continue. Ensuring the protection of key winter roost sites is strongly recommended.

Marshlands Conservancy

Rye, Westchester County

155 acres
0-50' elevation

40.9530°N
73.7019°W

IBA Criteria Met

Criterion	*Species*	*Data*	*Season*	*Source*
Species at Risk	Seaside Sparrow	1 ind. in 1999, 1 in 1997, 1 in 1995	Breeding	NY Natural Heritage Biodiversity Databases

Description: This site is an oasis for wildlife located on Long Island Sound in Westchester County. The site consists of diverse habitats, including mowed fields, forests, and extensive salt marshes; it is owned by the Westchester County Department of Parks, Recreation, and Conservation.

Birds: This site contains the largest remaining mainland salt marsh in the state, and supports at-risk species including the Common Loon, American Bittern, Osprey, Bald Eagle, Northern Harrier, Sharp-shinned Hawk, Cooper's Hawk, Golden Eagle, Peregrine Falcon, Clapper Rails, King Rail, American Woodcock, Roseate Tern, Common Tern, Least Tern, Black Skimmer, Yellow-breasted Chat, Saltmarsh Sharp-tailed Sparrows, and Seaside Sparrow. The site hosts a remarkable diversity of birds; close to 300 species have been recorded.

Conservation: This site is listed in the 2002 Open Space Conservation Plan as a priority site under the project name Westchester Marine Corridor. More inventories and monitoring of at-risk species are needed to better understand bird use and determine if this site continues to meet IBA criteria. A major concern is non-point source pollution and siltation (which will eventually necessitate dredging) from the Blind Brook watershed. Increased levels of boating activity originating in Milton Harbor have resulted in some instances of disturbance to breeding birds by kayakers, canoers, and illegal landing parties.

Mongaup Valley Wildlife Management Area

Multiple municipalities, Sullivan and Orange Counties

12,000 acres
490-1,400' elevation

41.5303°N
74.7673°W

IBA Criteria Met

Criterion	*Species*	*Data*	*Season*	*Source*
Species at Risk	Bald Eagle	4 pairs nesting, over 100 ind. in winter	Breeding and winter	Pete Nye pers. comm. 2004

Description: This site includes a series of reservoirs and streams flowing south from Mongaup Valley to the Delaware River, including the Mongaup Falls Reservoir and the Rio Reservoir. Water releases by the Orange and Rockland County Utilities keep the reservoirs and river from freezing completely, thereby providing a steady food supply for wintering eagles. NYS DEC administers the site and has built a viewing blind for watching eagles.

Birds: Mongaup supports one of the largest concentrations of wintering Bald Eagles in the state (numbers have exceeded 100 individuals), and several breeding pairs.

Conservation: This site is listed in the 2002 Open Space Conservation Plan as a priority site under the project name Mongaup Valley Wildlife Management Area. Portions of this site have been designated as a state Bird Conservation Area. The property is not open for general public use from December 1st through March 31st; however, eagle viewing is allowed from a blind or vehicle. Disturbance to nesting and wintering eagles should be monitored. The reservoirs and river should be kept from freezing completely to provide a foraging area for wintering eagles.

Northern Shawangunk Mountains

Multiple municipalities, Ulster County

90,000 acres
100-2,290' elevation

41.7245°N
74.2412°W

IBA Criteria Met

Criterion	*Species*	*Data*	*Season*	*Source*
Species at Risk	Wood Thrush	Extent of habitat and other data strongly suggest that the threshold is being met [1]; 2CO, 6PR, 3PO [2]	Breeding	1. Technical Committee consensus; 2. NY BBA 2000
Species at Risk	Prairie Warbler	Extent of habitat and other data strongly suggest that the threshold is being met [1]; 2CO, 5PR, 1PO [2]	Breeding	1. Technical Committee consensus; 2. NY BBA 2000
Species at Risk	Cerulean Warbler	Extent of habitat and other data strongly suggest that the threshold is being met [1]; 1CO, 1PR, 1PO [2]	Breeding	1. Technical Committee consensus; 2. NY BBA 2000
Species at Risk	Worm-eating Warbler	Extent of habitat and other data strongly suggest that the threshold is being met [1]; 2CO, 1PR, 1PO [2]	Breeding	1. Technical Committee consensus; 2. NY BBA 2000
Responsibility Species Assemblage-Forest	Sharp-shinned Hawk, Black-billed Cuckoo, Northern Flicker, Eastern Wood-Pewee, Least Flycatcher, Yellow-throated Vireo, Blue-gray Gnatcatcher, Wood Thrush, Black-throated Blue Warbler, Cerulean Warbler, Black-and-white Warbler, Worm-eating Warbler, Louisiana Waterthrush, Canada Warbler, Scarlet Tanager, Rose-breasted Grosbeak	Breed	Breeding	NY BBA 2000

Description: The northern Shawangunks represent a relatively intact landscape located between the Hudson River and the Catskill Mountains. The northern Shawangunks are bordered by the Wallkill River and Shawangunk Kill on the southeast, Route 52 on the southwest and Roundout Creek on the north. The area is a glacially scoured mountain range of white quartzite conglomerate rock with a patchy layer of shallow soils. The site has one of the most unusual pine barrens in the northeast and includes a largely unfragmented chestnut-oak forest. Wetlands exist in rock depressions, and a few deep lakes have formed in eroded faults and fractures. The site contains a globally rare 2,000-acre dwarf pine ridge community. The area supports 27 rare plants, nine rare insects, and three other rare natural communities. According to the NY GAP land cover data, approximately 85% of the site is forested, and includes Appalachian oak-pine, deciduous wetland, evergreen northern hardwood, evergreen plantation, evergreen wetland, oak, pitch pine-oak, successional hardwood, and sugar maple mesic forests. Agriculture is an active, but declining industry in the areas below the ridge. A mix of private, municipal, state, and non-governmental conservation agencies own the site, which includes over 12,000 acres administered by NYS OPRHP (Minnewaska State Park), 6,600 acres owned by the Mohonk Preserve, and 4,600 acres managed by The Nature Conservancy.

View from Sam's Point on the Northern Shawangunk Ridge

Birds: This site supports an exceptional example of a characteristic higher elevation forest bird community with particularly good representation of a pine woods community. Characteristic forest bird species breeding include the Sharp-shinned Hawk, Black-billed Cuckoo, Northern Flicker, Eastern Wood-Pewee, Least Flycatcher, Yellow-throated Vireo, Common Raven, Winter Wren, Blue-gray Gnatcatcher, Hermit Thrush, Wood Thrush, Black-throated Blue Warbler, Pine Warbler, Cerulean Warbler, Black-and-white Warbler, Worm-eating Warbler, Louisiana Waterthrush, Canada Warbler, Scarlet Tanager, and Rose-breasted Grosbeak. There are also breeding characteristic shrub species inhabiting this site including the Whip-poor-will, Golden-winged Warbler, Prairie Warbler, and Eastern Towhee, among others. Peregrine Falcons currently breed in one location on a cliff.

Conservation: This site is listed in the 2002 Open Space Conservation Plan as a priority site under the project name Shawangunk Mountains. With low levels of fragmentation and a high level of biological diversity, the northern Shawangunks is one of The Nature Conservancy's highest priorities for protection in the northeast. The lack of fire is a primary threat for this fire-adapted ecosystem, altering the structure of vegetation communities. Scattered residential development has a cumulative effect by fragmenting natural communities, thus threatening their integrity and limiting management options. Permanent protection and stewardship of private portions of the site are needed to prevent fragmentation resulting from development. Options include public or land trust acquisition, purchase of conservation easements, and sustainable forestry agreements. Approximately 550,000 people visit the area annually, generating $10-15 million in local economic activity. Rock climbers have the potential to impact cliff and talus communities, and disturb species such as Peregrine Falcons. There is an ongoing land stewardship and protection program for the area spearheaded by The Nature Conservancy. The Shawangunk Ridge Biodiversity Partnership, made up of public and private conservation agencies, has an active ecological research program; during the first round of IBA site identifications, this site was recognized under the research criterion. More research to document the importance of the area to at-risk species is needed.

Pepacton Reservoir

Multiple municipalities, Delaware County

5,000 acres
1,180-1,450' elevation

42.0882°N
74.8706°W

IBA Criteria Met

Criterion	Species	Data	Season	Source
Species at Risk	Bald Eagle	2 pairs in 2002, 1 in 1990-2001	Breeding	NY Natural Heritage Biodiversity Databases

Description: One of the largest and most pristine water bodies in central NY. The site serves as a public water supply and is owned by the City of New York. Surrounding lands are generally undisturbed beech/oak/maple forest, much of which is protected as a buffer zone. The southern portion of the reservoir is within the Catskill Park.

Birds: This site is important for nesting and wintering Bald Eagles. There has been at least one pair nesting here since 1990; currently there are two pairs. The area regularly hosts up to eight individual Bald Eagles during the winter. The reservoir also serves as a stopover site for waterbirds, including Common Loons and Ospreys, and the surrounding woodlands host a diversity of migratory and breeding land birds.

Conservation: This site is listed in the 2002 Open Space Conservation Plan as a priority site under the project name New York City Watershed Lands. Non-point source pollution from agricultural runoff and sewage upstream are a concern. A recently enacted watershed agreement provides funds to correct sewage problems and acquire lands. Access to the area is limited. Hunting and fishing are allowed by permit only, and powerboats are prohibited.

Rensselaer Forest Tract

Multiple municipalities, Rensselaer County

100,000 acres
490-2,075' elevation

42.7081°N
73.4448°W

IBA Criteria Met

Criterion	*Species*	*Data*	*Season*	*Source*
Species at Risk	Cooper's Hawk	6+ pairs	Breeding	Paul Connor pers. comm. 2004
Species at Risk	Northern Goshawk	Estimated 5+ pairs	Breeding	Paul Connor pers. comm. 2004
Species at Risk	Red-shouldered Hawk	Estimated 12-15 pairs	Breeding	Paul Connor pers. comm. 2004
Species at Risk	Wood Thrush	Estimated 15+ pairs	Breeding	Paul Connor pers. comm. 2004
Species at Risk	Canada Warbler	Estimated 70+ pairs	Breeding	Paul Connor pers. comm. 2004
Responsibility Species Assemblage-Forest	Ruffed Grouse, Sharp-shinned Hawk, Black-billed Cuckoo, Yellow-bellied Sapsucker, Eastern Wood-Pewee, Least Flycatcher, Great Crested Flycatcher, Blue-headed Vireo, Veery, Wood Thrush, Chestnut-sided Warbler, Black-throated Blue Warbler, Black-throated Green Warbler, Blackburnian Warbler, Blackpoll Warbler, Cerulean Warbler, Black-and-white Warbler, American Redstart, Ovenbird, Canada Warbler, Scarlet Tanager, Rose-breasted Grosbeak, Baltimore Oriole, Purple Finch	Breed	Breeding	NY BBA 2000

Description: A large, relatively unfragmented forest stretching approximately 12 miles from Grafton Lakes State Park in the north to Cherry Plain State Park and Capital District Wildlife Management Area in the south. The site includes two state parks administered by NYS OPRHP, the Capital District Wildlife Management Area administered by NYS DEC, and private lands. The site is situated on the elevated Rensselaer Plateau, which is cooler and more thickly wooded than the adjacent lowlands. Conifers are abundant. The climate, plus the shallow, poorly drained soils, favors an Adirondack-like forest. Wetlands abound, especially bogs and fens, and both natural and artificial ponds are common. Old foundations and other archeological evidence indicate past efforts at farming and charcoal production. According to the NY GAP land cover data, approximately 95% of the site is forested, and includes Appalachian oak-pine, deciduous wetland, evergreen northern hardwood, evergreen plantation, oak, successional hardwood, and sugar maple mesic forests.

Birds: The site supports a great abundance and diversity of forest breeders, including many at-risk species. At-risk species breeding at the site include the American Black Duck (at least six pairs), Pied-billed Grebe, American Bittern (use cattail wetlands in spring and might nest here), Sharp-shinned Hawk (at least five pairs), Broad-winged Hawk, Olive-sided Flycatcher (possible breeder), Willow Flycatcher (probable breeder), Common Raven, Winter Wren (at least 50), Golden-crowned Kinglet, Swainson's Thrush, Wood Thrush (15-30 pairs), Blue-winged Warbler (probable breeder), Prairie Warbler (possible breeder), and Cerulean Warbler (possible breeder). Characteristic forest breeders include the Ruffed Grouse, Black-billed Cuckoo, Yellow-bellied Sapsucker, Eastern Wood-Pewee, Least Flycatcher, Great Crested Flycatcher, Blue-headed Vireo, Veery, Chestnut-sided Warbler, Black-throated Blue Warbler, Black-throated Green Warbler, Blackburnian Warbler, Blackpoll Warbler, Black-and-white Warbler, American Redstart, Ovenbird, Canada Warbler, Scarlet Tanager, Rose-breasted Grosbeak, Baltimore Oriole, and Purple Finch. The Red Crossbill, White-winged Crossbill, Pine Siskin, and Evening Grosbeak are also found here. At least 18 species of warblers breed here, including the Nashville, Magnolia, Black-throated Blue, Yellow-rumped, Black-throated Green, Blackburnian, Mourning, and Chestnut-sided. Bicknell's Thrushes use the area during migration.

Grafton Lakes State Park

Conservation: This site is listed in the 2002 Open Space Conservation Plan as a priority site under the project name Rensselaer Plateau. This site is under development pressure as the Albany capital district expands, and forest fragmentation is a threat. Permanent protection of this site should be pursued via public acquisition and conservation easements. Working forest easements that promote sustainable forestry may provide habitat for species that require successional and disturbed forests. Additional negative impacts of development on wetlands could include sewage and storm water runoff. Timber managers should avoid logging in the vicinity of hawk nests. Some ponds have been drained or lowered resulting in less wetland habitat. Inventory and monitoring of breeding birds, especially at-risk species, should continue.

Rockefeller State Park Preserve and Forest

Mount Pleasant, Westchester County

1,400 acres
200-625' elevation

41.1113°N
73.8310°W

IBA Criteria Met

Criterion	*Species*	*Data*	*Season*	*Source*
Species at Risk	Wood Thrush	32-39 singing males 1996-2003	Breeding	Richard Nelson pers. comm. 2004
Species at Risk	Worm-eating Warbler	7-11 singing males	Breeding	Richard Nelson pers. comm. 2004

Description: This state park, formerly part of the Rockefeller Estate, together with privately owned adjacent lands, forms one of the largest undeveloped tracts of land in Westchester County. The site is located in the drainage basin of the Pocantico River and includes rolling hills and tributaries. Habitats include woodlands (mixed deciduous and hemlock), fields (some cut every year), wooded and shrubby wetland, wet meadow, and a 24-acre lake.

Birds: The mix of habitats supports a diversity of species. At-risk species breeding here include the Wood Thrush, Blue-winged Warbler (two or three singing males), Worm-eating Warbler, and Kentucky Warbler (one singing male). Surveys of a limited area of the park have documented three to eight singing male American Redstarts.

Conservation: Deer browsing and cowbird parasitism are thought to be significant threats to the habitat and to reproductive success of the birds. A research project has been proposed to study the effects of deer browsing on the understory. Adjacent lands will be added to the preserve as they become available. An additional 288 acres on Buttermilk Hill was recently added to the preserve.

Schodack Island State Park

Multiple municipalities, Columbia, Greene, and Rensselaer Counties

1,500 acres
0-85' elevation

42.4814°N
73.7787°W

IBA Criteria Met

Criterion	*Species*	*Data*	*Season*	*Source*
Species at Risk	Bald Eagle	Present	Breeding	NY Natural Heritage Biodiversity Databases
Species at Risk	Wood Thrush	20+ pairs	Breeding	Jane Graves pers. comm. 2004
Species at Risk	Cerulean Warbler	18 ind. in 1997	Breeding	Cerulean Warbler Atlas Project
Congregations-Wading Birds	Great Blue Heron	50+ pairs	Breeding	Ray Perry pers. comm. 2004

Wood Thrush

Description: Schodack is a narrow six-mile-long forested island in the Hudson River just south of Albany. The site supports day-use activities with a boat launch, eight miles of multi-use trails, and a bike trail. The site is administered by NYS OPRHP and US Army Corps of Engineers.

Birds: The area supported at least 18 Cerulean Warblers in 1997 and has been used by this species since at least 1965. Ospreys and Bald Eagles regularly use the area for roosting and perching. There is a Great Blue Heron rookery on the island with over 50 pairs.

Conservation: This site is listed in the 2002 Open Space Conservation Plan as a priority site under the project name Hudson River Corridor Estuary and Greenway Trail. A portion of the park has been designated a state Bird Conservation Area (BCA). The BCA is within a Department of State (DOS) Significant Coastal Fish and Wildlife Habitat and a DOS Significant Scenic Area. The BCA boundaries coincide with the designated Hudson River Estuarine Sanctuary boundaries. This area was previously relatively inaccessible and undeveloped, but a bridge to the island, a small boat launch, picnic area, and camping area were recently developed. The southern portion of the island where Bald Eagles regularly roost is being managed by NYS OPRHP in consultation with NYS DEC's Endangered Species Unit. Potential disturbance of eagles by dredge spoil dumping is being monitored. Inventory and monitoring, particularly of breeding Cerulean Warblers, are needed.

Shawangunk Grasslands

Shawangunk, Ulster County

530 acres
295-390' elevation

41.6381°N
74.2077°W

IBA Criteria Met

Criterion	*Species*	*Data*	*Season*	*Source*
Species at Risk	Northern Harrier	8 ind. in 2002, 7 in 2000, 4 in 1997, 15 in 1995, 36 in 1993	Winter	NY Natural Heritage Biodiversity Databases
Species at Risk	Short-eared Owl	6 ind. in 2002, 9 in 2000, 8 in 1999, 10 in 1997, 21 in 1995, 2 in 1993	Winter	NY Natural Heritage Biodiversity Databases
Species at Risk	Grasshopper Sparrow	5+ pairs	Breeding	Steve Kahl pers. comm. 2004
Responsibility Species Assemblage-Grassland	American Kestrel, Henslow's Sparrow, and Bobolink	Breed (no confirmed breeding of Henslow's in recent years, however, individuals have been observed)	Breeding	United States Fish and Wildlife Service 1998, 2001-2003

Description: The Shawangunk Grasslands National Wildlife Refuge, formerly known as the "Galeville Military Airport" is located in the Town of Shawangunk, Ulster County. In 1999, the General Services Administration transferred 566 acres to the U.S. Fish and Wildlife Service (USFWS) to create a new national wildlife refuge. The refuge comprises open grassland fields with a level or gently rolling topography, and deciduous woods along property boundaries. According to the NY GAP land cover data, approximately 25% of the site is open habitat, and includes cropland and old field/pasture land. Approximately 400 acres is actively managed as open field or grassland habitat and is dominated by Kentucky bluegrass (*Poa pratensis*). The site contains significant wetlands (38% of the area) scattered throughout the site. Several rare or uncommon plants grow on the refuge; species documented include Frank's sedge (*Carex frankii*), small-flowered agrimony (*Agrimonia parviflora*), purple milkweed (*Asclepias purpurascens*), small white aster (*Aster vimineus*), Bush's sedge (*Carex bushii*), coontail (*Ceratophyllum echinatum*), and watermeal (*Wolffia brazilinsis*). Although the surrounding landscape is rural, housing developments border three sides of the refuge. The southern side of the refuge borders a large horse farm.

Birds: The refuge is an important breeding and wintering area for grassland birds—one of few in the downstate region. It is also an important wintering area for raptors, including Northern Harriers (36 in 1993), Red-tailed Hawks, Rough-legged Hawks, American Kestrels, Short-eared Owls (many individuals present each winter, including 21 in 1995), and Northern Shrikes. Refuge winter raptor surveys frequently document seven to nine Short-eared Owls and 12 to 17 Northern Harriers (USFWS 2003, unpublished data). The John Burroughs Natural History Society (1969) reported a maximum of 21 Short-eared Owls, and Askildsen (1993) reported a maximum of 36 Northern Harriers. The site also supports a diverse grassland bird breeding community, including Northern Harriers, Upland Sandpipers, Vesper Sparrows, Savannah Sparrows, Grasshopper Sparrows, Henslow's Sparrows, Bobolinks, and Eastern Meadowlarks. The refuge conducts annual point-count surveys of breeding grassland birds in cooperation with the John Burroughs Natural History Society. In addition, the refuge also provides important habitat for migrant grassland birds in spring and fall. Northern Harriers migrating along the Shawangunk Mountains often stop at the refuge to rest and forage. Migrant Short-eared Owls arrive at the refuge in early November and depart in late April. Flocks of up to 100 Bobolinks gather at the refuge in August and September, and flocks of up to 50 Eastern Meadowlarks can be found in April, October, and November. Up to 19 Vesper Sparrows have been counted at the refuge in October (Kahl, USFWS 2001, personal observation).

Conservation: This site is listed in the 2002 Open Space Conservation Plan as a priority site under the project name Galeville Grasslands. This site is managed to maintain the dominance of grasses in the fields. Without frequent management, natural succession would shift that dominance to broadleaf herbaceous plants and shrubs such as goldenrod (*Solidago spp*), purple loosestrife (*Lythrum salicaria*), and gray dogwood (*Cornus racemosa*), causing the refuge to lose its habitat for grassland-dependent birds. Mowing is now the primary technique to halt succession. Future management techniques may include haying, grazing, discing, re-vegetating, applying herbicides, and prescribed burning. Monitoring of breeding grassland birds should continue.

Stissing Ridge

Multiple municipalities, Dutchess and Columbia Counties

11,400 acres
295-1,400' elevation

41.9549°N
73.7087°W

IBA Criteria Met

Criterion	*Species*	*Data*	*Season*	*Source*
Responsibility Species Assemblage-Forest	Black-billed Cuckoo, Eastern Wood-Pewee, Wood Thrush, Rose-breasted Grosbeak, Baltimore Oriole	Breed	Breeding	NY BBA 2000

Description: This site includes the relatively intact forest surrounding Stissing Mountain. Dominant trees of the forest community include red oak, ash, and beech, but the bases of the steep slopes tend towards a hemlock-mixed hardwood system. The site is surrounded by agricultural land. According to the NY GAP land cover data, approximately 90% of the site is forested, and includes Appalachian oak-pine, deciduous wetland, evergreen northern hardwood, evergreen plantation, oak, and sugar maple mesic forests. The Nature Conservancy owns Stissing Mountain Preserve (507 acres), but adjoining parcels are privately owned.

Birds: The site supports a characteristic community of forest breeders including the Black-billed Cuckoo, Barred Owl, Eastern Wood-Pewee, Acadian Flycatcher, Brown Creeper, Winter Wren, Hermit Thrush, Wood Thrush, Black-throated Blue Warbler, Canada Warbler, Rose-breasted Grosbeak, and Baltimore Oriole. A pair of Golden Eagles has regularly wintered here and there is a possibility that the birds may attempt nesting.

Conservation: Permanent protection and stewardship of private portions of the site are needed to prevent fragmentation resulting from development. Options include public or land trust acquisition, purchase of conservation easements, and sustainable forestry agreements. There is concern that recreational visitors, including birders, could disturb the eagles. As of 1993, the NYS DEC has posted the southern boundaries of the area and sought cooperation from The Nature Conservancy and local bird clubs to protect the birds from disturbance.

Stockport Flats

Multiple municipalities, Columbia and Greene Counties

2,150 acres
0-100' elevation

42.3123°N
73.7780°W

IBA Criteria Met

Criterion	*Species*	*Data*	*Season*	*Source*
Species at Risk	Least Bittern	20 pairs estimated 2000	Breeding	NY Natural Heritage Biodiversity Databases

Description: This marsh complex is centered around the confluence of Stockport Creek and the Hudson River. The Hudson River and marsh are subject to tidal flows. The tidal marsh contains a variety of habitats. The aquatic bed system includes water celery (*Vallisneria americana*), wild rice (*Zizania aquatica*), and narrowleaf cattail (*Typha angustifolia*). The upland banks are thickly forested with floodplain species and dense undergrowth. A number of rare plants occur at the site, including heart leaf plantain (*Plantago cordata*), estuary beggar ticks (*Bidens bidentoides*), kidneyleaf mud-plantain (*Heteranthera reniformis*), and spongy arrowhead (*Sagittaria calycina* var. *spongiosa*). Most of the site is administered by NYS DEC, NYS OPRHP, and the New York State Office of General Services, but there are several small, privately owned parcels as well.

Birds: The freshwater tidal marsh supports the Pied-billed Grebe (migration), Least Bittern (breeding), Bald Eagle (breeding and wintering), Northern Harrier (migration), Least Bittern (breeding), Virginia Rail, and Marsh Wren. A large variety of waterfowl use the site during spring and fall. Thousands of swallows stage here in late summer. Belted Kingfishers and Bank Swallows nest in the sand cliffs on the southwest shore of Stockport Middle Ground Island.

Conservation: This site is listed in the 2002 Open Space Conservation Plan as a priority site under the project name Hudson River Corridor Estuary and Greenway Trail. Purple loosestrife (*Lythrum salicaria*) grows in portions of the complex and should be controlled. Other potential threats are dredge operations in the river that would result in dumping in the marsh. The site is part of the Hudson River National Estuarine Research Reserve and has been identified as a Significant Coastal Habitat by the NYS Department of State. Better inventory and monitoring of wetland birds are needed.

Tivoli Bays

Red Hook and Saugerties, Dutchess and Ulster Counties

1,800 acres 42.0320°N
0-200' elevation 73.9185°W

IBA Criteria Met

Criterion	*Species*	*Data*	*Season*	*Source*
Species at Risk	Least Bittern	1 ind. in 2000, 4 in 1998, 1 in 1994	Breeding	NY Natural Heritage Biodiversity Databases

Description: The Tivoli Bays are the only large fresh-tidal marshes on the Hudson River primarily bordered by undeveloped forest. Tivoli North Bay is a freshwater tidal marsh dominated by cattail; Tivoli South Bay includes vegetated tidal shallows (mudflats at low tide). The site is an important spawning and nursery ground for a variety of anadromous and freshwater fish species. Map turtle (*Graptemys geographica*) and American brook lamprey (*Lampetra appendix*), both rare in New York, occur at the site. The site hosts a number of rare plant species including estuary beggar ticks (*Bidens bidentoides* and *B. hyperborea*), ovate spikerush (*Eleocharis ovata*), Eaton's bur-marigold (*Bidens eatoni*), winged monkey flower (*Mimulus alatus*), and swamp lousewort (*Pedicularis lanceolata*). The site is primarily administered by NYS DEC and the New York State Office of General Services, but contains several small, privately owned parcels as well.

Birds: Species documented as breeders include the Least Bittern (estimates from 1970s suggested dozens of pairs, recent numbers are probably similar), Virginia Rail, Sora (apparently bred until mid-1970s), Common Moorhen (apparently bred until mid-1970s or 1980s), and Marsh Wren (estimates from 1970s suggested a thousand individuals, recent numbers are probably similar). The site is one of the last known areas for King Rails in the state. The area is used by large flocks of dabbling ducks in the spring and fall, with good numbers of American Black Ducks (peak estimates of 100-200 individuals). Bald Eagles forage at this site year round, and Ospreys and Northern Harriers regularly forage during migration. Large fall post-breeding concentrations of swallows possibly reach 10,000 individuals.

Conservation: This site is listed in the 2002 Open Space Conservation Plan as a priority site under the project name Hudson River Corridor Estuary and Greenway Trail. Common reed (*Phragmites australis*) covers approximately one percent of North Bay and is spreading; management is being debated but is not currently practiced. Eurasian water chestnut (*Trapa natans*) dominates South Bay and is very minor in North Bay; no control is currently practiced due to concerns about non-target impacts of herbicides. Purple loosestrife (*Lythrum salicaria*) has been declining. Potential pollution leaching from a nearby landfill (now closed) should be monitored. Better inventory and monitoring of the marsh birds are needed.

Marsh Wren

Upper Delaware River

Multiple municipalities, Sullivan, Orange, Delaware, and Broome Counties

29,000 acres
415-1,375' elevation

41.6490°N
75.0213°W

IBA Criteria Met

Criterion	*Species*	*Data*	*Season*	*Source*
Species at Risk	Bald Eagle	5 pairs in 2004, 6 in 2003, 5 in 2002, 5 in 2001, 4 in 2000, 4 in 1999, 2 in 1998, 1 in 1997-1993; 28 ind. in the winter of 1998, 77 in 1999, 145 in 2000, 27 in 2001, 109 in 2002, 41 in 2003, and 28 in 2004	Breeding and winter	NYS DEC Bald Eagle Report 2004, Eagle Institute, and Sullivan County Audubon Society 2002
Congregations-Individual Species	Bald Eagle	Around 5% of state breeding population, and between 5-40% of state wintering population	Breeding and winter	NYS DEC Bald Eagle Report 2004, Eagle Institute, and Sullivan County Audubon Society 2002

Description: The Upper Delaware River includes a 73-mile stretch of the longest undammed river in the northeast. It is managed by the National Park Service as a Scenic and Recreational River; however, federal ownership includes only 31 acres of riparian land. The Upper Delaware separates New York and Pennsylvania. Five counties and 15 towns form its borders. Diverse habitats are adjacent to the river, including riparian, wooded, open field, and agricultural lands. Also found along the river are residential and commercial developments, a state highway (designated as a scenic highway in 2004), smaller rural roads, and an active railroad line. The river varies from white water to calm pools, eddies, and shallow ripples. Some rock ledges and sloping lands rise up about a quarter mile from each shoreline. There are several tributaries and waterfalls.

Birds: This is one of the most important sites in the state for Bald Eagles. Several locations on the river remain free of ice in winter, and more than 100 eagles congregate there for food. Several significant roost sites have been identified as well. The site supports almost 10% of the state's breeding population, and up to 1/3 of the state's wintering population of Bald Eagles. It is also important for migrating and wintering waterfowl.

Conservation: Audubon Pennsylvania also has designated the Upper Delaware River as an IBA. A River Management Plan for the National Park Service is in place for this site. The Upper Delaware Council, composed of representatives of the 15 river corridor towns, five counties, and resource management agencies in the two states, oversees activity within the river corridor. Development pressures have been increasing, and growth should be managed in a way that protects the river's habitats, especially for eagles. However, zoning and planning regulations vary by town and may not be consistent with habitat protection needs. Numerous environmental organizations, including the Eagle Institute, Sullivan County Audubon Society, and the Delaware Highlands Conservancy, act as advocacy and educational entities in the local communities. The Upper Delaware is a tremendous economic asset to the communities within and adjacent to the river corridor. Fishing, hunting, birding, boating, hiking, and other types of outdoor recreation on the Upper Delaware are key components of the region's economy. Logging along the Delaware River corridor (including upstream and the east and west branches of the river) are also a concern, particularly in regard to loss of habitat. Significant, dedicated set-asides of remaining Delaware corridor habitats will be required to ensure perpetuation of sensitive wildlife. Data collected by the Eagle Institute show over 4,000 visitors coming to the region every winter for eagle watching, which contributes to the economy during a slower, non-traditional tourism season. The Sullivan County Visitors Association heavily promotes this activity. More monitoring of waterfowl and other at-risk species is needed to better understand their use of this area. Specific needs include the identification of critical eagle roost areas and potential breeding sites, and continued monitoring of migrant and nest fledglings using satellite-radio tagging.

Vischer Ferry Nature and Historic Preserve

Clifton Park, Saratoga County

465 acres
155-190' elevation

42.7892°N
73.7961°W

IBA Criteria Met

Criterion	*Species*	*Data*	*Season*	*Source*
Congregations-Wading Birds	Mixed species	Max. of 120 ind. 1991	Fall	Able, K.P. 1991

Pied-billed Grebe

Description: This site is a large freshwater wetland along the northern edge of the Mohawk River in Saratoga County. It is popular with birders because of its easily observed marshes and ponds. The site is primarily administered by the New York State Thruway Authority, but is leased and managed by the Town of Clifton Park, with adjoining acreage owned by the state.

Birds: This site contains important wetland habitat with cattail marshes that support at-risk species, including the American Black Duck (migration), Pied-billed Grebe (migration), American Bittern (breeding season), Least Bittern (probable breeder), Osprey (migration), Northern Harrier (migration), Sharp-shinned Hawk (migration), Cooper's Hawk (migration), Common Nighthawk (migration), Willow Flycatcher (migration), Wood Thrush (probable breeder), Blue-winged Warbler (migration), Bay-breasted Warbler (migration), Canada Warbler (migration), and Rusty Blackbird (migration). This site is particularly important as a concentration area for post-breeding herons and egrets. In 1991, it hosted 102 Great Blue Herons, 17 Great Egrets, and one Black-crowned Night-Heron.

Conservation: Water level control is critically needed for this site to continue supporting the same numbers of wetland bird species it has in the past. The current master plan recommendations could increase recreational use, which might negatively impact this site's avifauna. Positive aspects of the master plan include suggestions to reduce the non-native vegetation encroaching on wetlands and acquire additional lands nearby. Plans for this site should explicitly recognize its importance to wetland birds. More inventories and monitoring of at-risk species and wading birds are needed to better understand bird use and determine whether this site continues to meet IBA criteria.

Ward Pound Ridge

Bedford, Lewisboro, Pound Ridge, Westchester County

17,000 acres
295-825' elevation

41.2496°N
73.5938°W

IBA Criteria Met

Criterion	Species	Data	Season	Source
Responsibility Species Assemblage-Forest	Broad-winged Hawk, Black-billed Cuckoo, Hairy Woodpecker, Northern Flicker, Eastern Wood-Pewee, Great Crested Flycatcher, Wood Thrush, Black-and-white Warbler, Scarlet Tanager, Rose-breasted, Grosbeak, Baltimore Oriole	Breed	Breeding	NY BBA 2000

Description: This area includes the largest park in Westchester County—Ward Pound Ridge Reservation (4,700 acres)—and surrounding forested areas. According to the NY GAP land cover data, approximately 80% of the site is forested, and includes Appalachian oak-pine, deciduous wetland, evergreen northern hardwood, oak, and sugar maple mesic forests. The park includes more than 35 miles of hiking trails, camping facilities, horse trails, a wildflower garden, and the Trailside Museum. The site supports significant populations of reptiles, including wood turtles and box turtles. A number of 18th and 19th century houses remain on the property. Portions of the park support natural communities typically found further north. A large part of the site is owned by the Westchester County Department of Parks, Recreation, and Conservation, but much of the land is under private ownership.

Birds: This site supports an exceptional regional bird community, representative of the hardwood forests of southern New England. Breeding species include the Broad-winged Hawk, Black-billed Cuckoo, Hairy Woodpecker, Northern Flicker, Eastern Wood-Pewee, Great Crested Flycatcher, Yellow-throated Vireo, Wood Thrush, Black-and-white Warbler, Worm-eating Warbler, Louisiana Waterthrush, Scarlet Tanager, Rose-breasted Grosbeak, and Baltimore Oriole. Additional forest breeders include the Ruffed Grouse, Northern Goshawk, Winter Wren, Blue-winged Warbler, and Black-throated Green Warbler.

Ward Pound Ridge

Conservation: Development and fragmentation of portions of this site outside of Ward Pound Ridge Reservation are major concerns. Permanent protection and stewardship of private portions of the site are needed to prevent fragmentation resulting from development. Options include public or land trust acquisition, purchase of conservation easements, and sustainable forestry agreements. This site needs better monitoring of at-risk species.

Chapter 3.4
New York City and Long Island

Sunken Meadows State Park

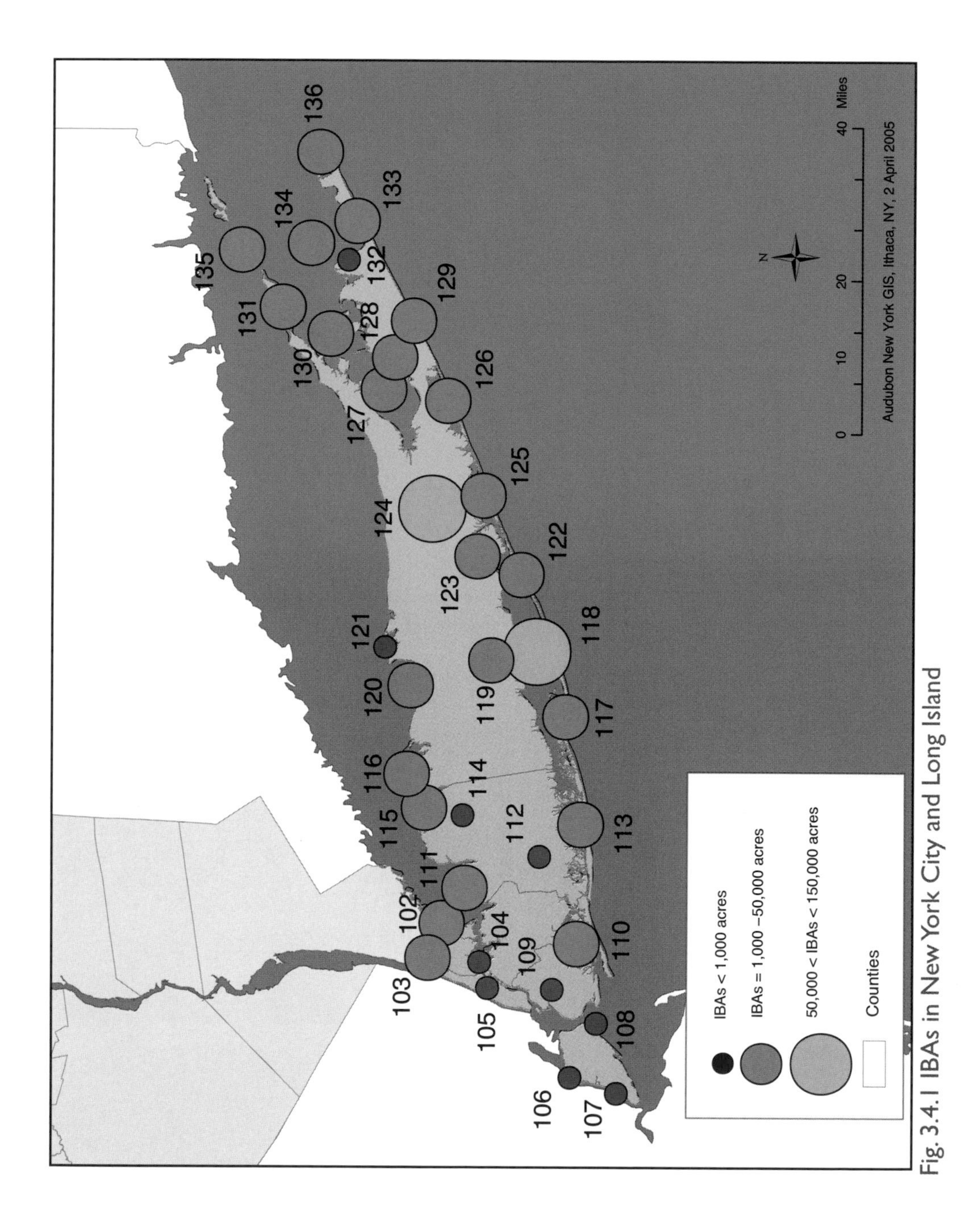

Fig. 3.4.1 IBAs in New York City and Long Island

Accabonac Harbor

East Hampton, Suffolk County

740 acres
0-35' elevation

41.0255°N
72.1424°W

IBA Criteria Met

Criterion	*Species*	*Data*	*Season*	*Source*
Species at Risk	Osprey	8 active nests	Breeding	Mike Wasilco pers. comm. 2004
Species at Risk	Piping Plover	5 pairs in 2004 [1]; 3 in 1999, 3 in 1998, 7 in 1997, 3 in 1996, 3 in 1995, 5 in 1994, 3 in 1993 [2]	Breeding	1. NY Natural Heritage Biodiversity Databases 2. Mike Wasilco pers. comm. 2004
Species at Risk	Least Tern	70 pairs in 1998, 352 in 1997, 567 in 1996, 235 in 1995, 326 in 1994, 731 in 1993	Breeding	NY Natural Heritage Biodiversity Databases

Description: Accabonac Harbor is one of the major undeveloped coastal wetland ecosystems on Long Island. It includes a shallow open water bay surrounded by extensive salt marshes, sand spits, spoil disposal areas, and small wooded islands. The harbor is surrounded by undeveloped woodlands, except for a residential development to the north. Bay scallops and hard clams are abundant in the harbor, which supports a regionally important recreational shellfishery. The harbor also serves as a nursery and feeding area for a variety of estuarine fish species. The Town of East Hampton owns the harbor and part of the shoreline; the rest is held by private landowners and The Nature Conservancy.

Birds: Waterfowl wintering here in 1995 included Canada Geese (3,035), White-winged Scoters (238), Long-tailed Ducks (119), and Red-breasted Mergansers (129). The site supports at-risk species including the Osprey (breeds, and breeders from surrounding area feed here), Northern Harrier (winters), Piping Plover (breeds), Least Tern (breeds), Short-eared Owl (winters), Saltmarsh Sharp-tailed Sparrow (breeds), and Seaside Sparrow (breeds).

Osprey

Conservation: This site is listed in the 2002 Open Space Conservation Plan as a priority site under the project name Peconic Pinelands Maritime Reserve Project. Water pollution, chemical contamination, oil spills, and sewage or storm-water runoff, will adversely affect the site's biological productivity. Activities that could lead to these problems should be prevented. Construction and maintenance of shoreline structures, including docks, piers, and bulkheads, should be avoided in undisturbed areas. Elimination of salt marsh and intertidal areas through excavation or filling could decrease available habitat for salt marsh breeding birds and should be prevented. Nesting Piping Plovers and Least Terns are highly susceptible to human disturbance and recreational activities near nesting areas, including boat landing, off-road vehicle use, and picnicking, should be minimized from April through mid-August. Fencing, beach closures, posting, beach warden patrols, and public education may be required in some areas. Monitoring of at-risk species and waterfowl numbers should continue.

Captree Island Vicinity

Babylon and Islip, Suffolk County

14,300 acres
0-10' elevation

40.6314°N
73.3153°W

IBA Criteria Met

Criterion	*Species*	*Data*	*Season*	*Source*
Species at Risk	Piping Plover	23 pairs in 2001, 21 in 2000, 16 in 1999, 8 in 1998, 14 in 1997, 12 in 1996, 15 in 1995, 8 in 1994, 10 in 1993	Breeding	NY Natural Heritage Biodiversity Databases
Species at Risk	American Oystercatcher	31 pairs in 1995	Breeding	NY Natural Heritage Biodiversity Databases
Species at Risk	Common Tern	185 pairs in 2002, 387 in 2001, 345 in 2000, 430 in 1999, 907 in 1998, 953 in 1997, 920 in 1996, 806 in 1995, 1,502 in 1994, 1,856 in 1993	Breeding	NY Natural Heritage Biodiversity Databases
Species at Risk	Least Tern	35 pairs in 2001, 78 in pairs 2000, 70 in 1999, 108 in 1998, 100 pairs in 1997, 62 in 1996, 253 in 1995, 209 in 1994, 160 in 1993	Breeding	NY Natural Heritage Biodiversity Databases
Species at Risk	Black Skimmer	28 pairs in 1999, 51 in 1998, 84 in 1997, 119 in 1996, 244 in 1995, 233 in 1994, 225 in 1993	Breeding	NY Natural Heritage Biodiversity Databases
Species at Risk	Seaside Sparrow	4+ pairs in 2001	Breeding	NY Natural Heritage Biodiversity Databases
Responsibility Species Assemblage-Wetland	American Black Duck, Glossy Ibis, Clapper Rail, Marsh Wren, Saltmarsh Sharp-tailed Sparrow, Seaside Sparrow	Breed	Breeding	NY BBA 2000
Congregations-Waterbirds	Roseate, Common, and Least Terns	85 ind. in 2002, 422 in 2001, 425 in 2000, 501 in 1999, 1,031 in 1998, 1,062 in 1997, 1,012 in 1996, 1,078 in 1995, 1,786 in 1994, 2,073 in 1993	Breeding	NY Natural Heritage Biodiversity Databases

Description: This site includes the barrier islands on the south shore of Long Island, and the islands and marshes on the bayside. Sandy beach and dune systems, natural salt marshes, and spoil islands are included. According to the NY GAP land cover data, approximately 20% of the site is salt marsh habitat. The site extends from the Nassau/Suffolk county line east to and including Captree Island and Robert Moses State Park. It includes the eastern end of Jones Beach Island and the western tip of Fire Island. The interior of the barrier island is bisected by a four-lane highway with associated heavily developed recreational areas and large parking areas. Ownership is a mix of public (Captree Island State Park, Gilgo State Park, and Robert Moses State Park, administered by New York State Office of Parks, Recreation and Historic Preservation [NYS OPRHP]), municipal, and private.

Birds: This site supports high numbers of wading birds during the breeding season: 125 pairs in 1993, 140 in 1992, 54 in 1991, 206 in 1990, 365 in 1989, 194 in 1988, 305 in 1987, 375 in 1986, 171 in 1985, 120 in 1984. Wading birds include Great Egrets (6 pairs in 1995, representing 1% of the state's coastal population), Snowy Egrets (10 pairs in 1995; 2% of state population), Little Blue Herons (5 pairs in 1995; 19% of state population), Tricolored Herons (10 pairs in 1995; 38% of state population), Black-crowned Night-Herons (75 pairs in 1995; 4% of state coastal population), and Glossy Ibis (80 pairs in 1995; 11% of state population). In recent years, the total number of wading birds has dropped to under 100 individuals. The site supports at-risk species, including Northern Harriers (breeds and migrant), Black Rails (one pair in 1997, the only known breeding location in the state), Piping Plovers (eight pairs in 1994; 4% of state breeding population), American Oystercatchers (31 pairs in 1995; 17% of state population), Herring Gulls (893 pairs in 1995; 8% of state population), Great Black-backed Gulls (68 pairs in 1995; 1% of state population), Roseate Terns (75 pairs in 1994; 5% of state population), Common Terns (2,000 pairs in 1994; 12% of state coastal population), Least Terns (200 pairs in 1994; 8% of state population), Black Skimmers (33 pairs in 1994; 6% of state population), Short-eared Owls (breeds), Horned Lark (breeds and migrant), Saltmarsh Sharp-tailed Sparrows, and Seaside Sparrows. Other salt marsh breeders include Clapper Rails and Willets. The area is also important for passerine migrants and raptors, particularly in the fall. The tidal area at Democrat Point at the western tip of Fire Island hosts a great diversity and abundance of shorebirds. This is one of the few sites in the state with regularly breeding Chuck-will's-widow.

Conservation: This site is listed in the 2002 Open Space Conservation Plan as a priority site under the project name Long Island South Shore Estuary Preserve. Portions of this site have been designated as a state Bird Conservation Area. Portions of the beaches are heavily used for recreation. Continue management efforts to eliminate or minimize human disturbance and intrusions into nesting colonies of terns and Piping Plovers on beaches during their critical nesting season (mid-April to August). Efforts could include fencing, beach closures, posting, beach warden patrols, and public education. In those colonies where predation is a significant problem, whether from pets, feral animals, or native species such as raccoons or gulls, predator control programs should be undertaken. Predator control at Cedar Beach, where predation resulted in the complete loss of a large tern and skimmer colony in 1995, should be a high priority. Protection of the full geographical extent of both current and recent historical nesting beaches should be sought as a means of ensuring the long-term survival of beach-nesting birds in this region. NYS OPRHP manages a Piping Plover and Least Tern protection program in cooperation with various agencies and interest groups. The New York State Department of Environmental Conservation (NYS DEC) coordinates an annual survey of Piping Plovers and colonial waterbirds.

Saltmarsh Sharp-tailed Sparrow

Carman's River Estuary

Brookhaven, Suffolk County

8,000 acres
0-115' elevation

40.7990°N
72.9125°W

IBA Criteria Met

Criterion	*Species*	*Data*	*Season*	*Source*
Species at Risk	Osprey	8 active nests	Breeding	Florence James (USFWS) pers. comm. 2004
Responsibility Species Assemblage-Wetland	American Black Duck, Clapper Rail, Virginia Rail, Marsh Wren, Saltmarsh Sharp-tailed Sparrow, Seaside Sparrow	Breed [1] and [2]	Breeding	1. Florence James (USFWS) pers. comm. 2004; 2. NY BBA 2000
Congregations-Waterfowl	Mixed species	3,000-4,000 individuals on average	Winter	Bob Parris (USFWS) pers. comm. 1997

Description: Situated on the south shore of Long Island, this site includes the Carman's River, a New York State designated Wild and Scenic River, and its estuary, as well as uplands composed of oak and pine barren vegetation (part of the Long Island Pine Barrens). The core protected portion of the area is the 2,550-acre Wertheim National Wildlife Refuge (NWR). According to the NY GAP land cover data, this site includes approximately 675 acres of salt marsh habitat. The estuary provides an important spawning and nursery area for an abundance of fish and other aquatic life and is one of only four known breeding sites in the state for the eastern mud turtle (*Kinosternon subrubrum*). The site is primarily owned by the United States Fish and Wildlife Service (USFWS) and Suffolk County Parks, and the rest is privately owned.

Birds: This site is important for breeding and wintering waterfowl (3,000-4,000 on average), including large numbers of American Black Ducks (60% of all waterfowl at the site) and Greater Scaup at the mouth of the river. Hooded and Common Mergansers winter further upriver, where the site provides open water in the winter when the bay freezes. The area also supports the largest breeding population of Wood Ducks on Long Island. Carman's River's marshes support breeding at-risk birds, including the American Black Duck (probably exceeds IBA threshold, but further data are needed), American Bittern, Least Bittern (probably exceeds IBA threshold, but further data are needed), Osprey, Bald Eagle (winters), Saltmarsh Sharp-tailed Sparrow (probably exceeds IBA threshold, but further data are needed), and Seaside Sparrow (probably exceeds IBA threshold, but further data are needed). Clapper Rails and Willets are also found here. During fall migration, the marshes support 5-10 shorebirds per acre, including the Semipalmated Plover, Greater and Lesser Yellowlegs, Semipalmated Sandpiper, Least Sandpiper, and Pectoral Sandpiper. Also, wading birds can be seen along the river and refuge marshes, including the Great Blue Heron, Great Egret, Snowy Egret, Little Blue Heron, and Glossy Ibis. Migrating Tree Swallows come to the marshes along the Carman's River in the last weeks of September to roost. The estimated flock size is many thousands, numbering in the tens of thousands on some nights. These marshes also provide important habitat for thousands of Red-winged Blackbirds. The banks are bordered mostly with common reed (*Phragmites australis*), with some common cattail (*Typha latifolia*) and other brackish tolerant species. The swallows primarily congregate in the marshes that are part of the Wertheim Wildlife Refuge.

Conservation: This site is listed in the 2002 Open Space Conservation Plan as a priority site under the project name Long Island South Shore Estuary Preserve. Although a large portion of the area is protected, there is great pressure for residential development on some parcels. The refuge has identified several inland holdings that warrant protection. These currently unprotected lands should be acquired, or easements negotiated. In some wetland areas the encroachment of common reed *(Phragmites australis)* and other invasive plants threaten the habitat and should be controlled. Common reed is actively managed within refuge impoundments to increase habitat for migratory birds and other fauna. An Osprey nest camera has been installed to allow online viewing. An open marsh water management restoration project is being implemented at Wertheim NWR's East Marsh. This project will study marsh dynamics, restore salt marsh hydrology, control invasive plants, create reservoirs for baitfish, and reduce mosquito breeding areas. Refuge personnel are currently working on restoring up to 50 acres of grassland by controlling invasive woody and broadleaf vegetation, maintaining current grasslands, and removing unused buildings. Also, a comprehensive conservation plan for the refuge has been completed, and staff are working on a habitat management plan that addresses deer management. These plans will set goals and objectives for managing the refuge for the benefit of wildlife for the next 15 to 20 years. Inventory and monitoring, particularly of at-risk species, should continue.

Central Park

Manhattan, New York County

840 acres
45-100' elevation

40.7824°N
73.9655°W

IBA Criteria Met

Criterion	*Species*	*Data*	*Season*	*Source*
Congregations-Migrant Landbirds	Mixed species	More than 270 species have been recorded in the park; common during spring migration to see 100 or more species in a day, including 25 species of warblers. It is estimated that many thousands of birds use the site during migration.	Migration	Nomination, NY City Audubon

Description: Central Park is a human-made conglomeration of woodlots, meadows, lakes, playing fields, playgrounds, roads, and developed structures within the borough of Manhattan in the heart of New York City. It serves as a migration stopover point along the Atlantic Flyway. The park represents rare habitat both within the city and along the highly urbanized northeastern stretch of the flyway.

Birds: The park's relatively large size and position as an island of forest and wetland habitat in the midst of a sea of nearly complete urbanization makes it particularly important as a stopover site along the Atlantic Flyway. More than 270 species have been recorded in the park and it is not uncommon in spring to see 100 or more species in one day. Often, over 25 of these species are warblers. It is estimated that many thousands of flycatchers, swallows, thrushes, vireos, warblers, sparrows, and other migrants regularly use the site during spring and fall migration. A hawk watch at the site has tallied over 8,000 hawks in the fall (12,486 in 1994). Also, the site regularly supports a diversity of waterfowl, including approximately 2,000 wintering Ruddy Ducks (January 1998) and Pied-billed Grebes.

Conservation: Park management resides with the New York City Department of Parks and Recreation and the private Central Park Conservancy (which raises 80% of park funds). Park changes and maintenance procedures are usually extensively reviewed by New York City Audubon, local community boards, other environmental groups, and active park users before implementation. The millions of visitors make human disturbance a great threat. Soil compaction and erosion and invasive plants stemming from human disturbance threaten this site. Large numbers of introduced bird species, especially European Starlings and House Sparrows, compete with native birds. Air pollution and insect infestation have had detrimental effects on evergreens in the park. The 1990s North Woods projects have been highly successful in restoring soil and vegetation, creating wildlife islands and a wildflower meadow, removing invasive exotics, and educating volunteers and the public. An ongoing project for The Ramble (a migratory bird hotspot in the park) is to define desired habitat types and species and implement a five-year plan to attain and maintain them. Due to an Asian long-horned beetle (*Anoplophora glabripennis*) infestation in the Hallet Nature Sanctuary, the sanctuary is now being restored with native shrubs and trees. Although Central Park has a rich history of use by birders and a database of bird sighting information, there is a need for a carefully designed, quantitative study of the numbers of migrants that use the park, how they use the various habitats, and what foods they rely on. During the first round of IBA site identifications, this site was recognized under the research criterion because a long-term monitoring project is based there.

Clay Pit Ponds State Park Preserve

Staten Island, Richmond County

260 acres
40-100' elevation

40.5406°N
74.2290°W

IBA Criteria Met

Criterion	*Species*	*Data*	*Season*	*Source*
Congregations-Migrant Landbirds	Mixed species	180 species, 57 neotropical migrants, 31 warblers have been documented at this site	Migration	Nomination, NY City Audubon

American Woodcock

Description: Clay Pit Ponds State Park Preserve is a 260-acre natural area administered by the NYS OPRHP located near the southwest shore of Staten Island. Once the site of a clay-mining operation, the park today contains a mixture of locally uncommon habitats such as wetlands, fields, sand barrens, spring-fed streams, and woodlands. As a terminal point for some northern and southern plant species, the area is also rich with a variety of flora and non-avian fauna.

Birds: Over 180 bird species have been recorded at this site, including 40 regularly occurring neotropical migrants and at-risk breeding species. Characteristic shrub/pine barrens species such as the Northern Bobwhite, American Woodcock, Whip-poor-will, Eastern Kingbird, Gray Catbird, Brown Thrasher, Blue-winged Warbler, Prairie Warbler, Eastern Towhee, and Field Sparrow can be regularly found here.

Conservation: In 1986 NYS OPRHP created the "Clay Pit Ponds State Park Preserve Management Plan," which addresses many threats to the site, including non-native flora/fauna and erosion. NYS OPRHP also provides educational programs, including bird walks. Park use is restricted to passive recreation.

Connetquot Estuary

Islip, Suffolk County

7,700 acres
0-55' elevation

40.7561°N
73.1521°W

IBA Criteria Met

Criterion	*Species*	*Data*	*Season*	*Source*
Responsibility Species Assemblage-Shrub	Northern Bobwhite, American Woodcock, Whip-poor-will, Eastern Kingbird, Gray Catbird, Brown Thrasher, Blue-winged Warbler, Prairie Warbler, Eastern Towhee, Field Sparrow	Breed	Breeding	NY BBA 2000

Description: This site is located within the Connetquot River watershed and includes Connetquot River State Park, Benton Bay, Heckscher State Park, and surrounding lands. Uplands include relatively large areas of pine barrens, oak-pine woodlands, and oak brush plains. Salt marsh/tidal creek wetlands are found near the mouth of river. According to the NY GAP land cover data, approximately 50% of this site is shrub habitat, which includes old/field pasture and pitch pine oak. The Connetquot is one of only four major rivers on Long Island, and supports one of the few wild brook trout (*Salvelinus fontinalis*) populations on Long Island. A number of rare plants occur in the area as well.

Birds: This is one of the largest areas of undeveloped pitch pine/scrub oak and general scrub habitat on eastern Long Island. It harbors significant populations of characteristic shrub/scrub species, including the Northern Bobwhite, American Woodcock, Whip-poor-will, Eastern Kingbird, Gray Catbird, Brown Thrasher, Blue-winged Warbler, Prairie Warbler, Eastern Towhee, and Field Sparrow. Small numbers of breeding Common Terns (35 pairs in 1997) and Least Terns (nine pairs in 1997) are present at Timber Point. Tidal wetland habitats also support breeding Willets, Marsh Wrens, and Saltmarsh Sharp-tailed Sparrows, and provide important foraging habitat for Snowy Egrets and Least Terns.

Conservation: This site is listed in the 2002 Open Space Conservation Plan as a priority site under the project name Long Island South Shore Estuary Preserve. Portions of this site have been designated as a state Bird Conservation Area. Much of the land within the Connetquot Estuary is under public ownership administered by NYS OPRHP. However, some wetlands and uplands within the area are not currently protected and should be given priority for acquisition. In 1999, the state acquired Benton Bay with NYS Clean Water/Clean Air Bond Act funds. Benton Bay is a 127-acre critical wetland located in Oakdale at the mouth of the Connetquot River between Brick Kiln and Ludlow Creeks. The parcel includes approximately 50 acres of tidal wetlands, 50 acres of freshwater wetlands, and 27 acres of associated upland areas. Habitat and recreation management in the nearby state parks should be carried out in accordance with the needs of the characteristic breeding birds of the area. Sites should be inventoried and monitored for at-risk species.

Heckscher State Park

Crane Neck to Misery Point

Brookhaven, Suffolk County

740 acres
0-32' elevation

40.9707°N
73.1074°W

IBA Criteria Met

Criterion	*Species*	*Data*	*Season*	*Source*
Species at Risk	Piping Plover	6 pairs in 1999, 4 in 1998, 4 in 1997, 3 in 1996, 2 in 1995, 1 in 1994, 1 in 1993	Breeding	NY Natural Heritage Biodiversity Databases
Species at Risk	Least Tern	7 pairs in 1999, 64 in 1998, 120 in 1997, 23 in 1996, 32 in 1995, 77 in 1994	Breeding	NY Natural Heritage Biodiversity Databases

Description: This site is located on Long Island Sound and encompasses the shoreline of Old Field peninsula from Crane Neck Point in the west to Mount Misery Point on the east side of Port Jefferson Harbor (including Flax Pond, a pristine tidal embayment). The shoreline consists of unvegetated sand/cobble beach strand, backed in places by steep bluffs. The site is owned by a mix of private, municipal, county, and state groups.

Birds: This site supports colonial breeding birds, including the Piping Plover, Common Tern (four individuals in 1997), and Least Tern. Historically, larger numbers of all of these species occurred at the site. For example, eight pairs of Piping Plovers occurred in 1985, 83 pairs of Common Terns in 1989, and over 200 pairs of Least Terns in 1984.

Conservation: This site is listed in the 2002 Open Space Conservation Plan as a priority site under the project name Long Island Sound Coastal Area. Limited public vehicular access minimizes the threat of human disturbance at this site. Old Field Beach, however, continues to be heavily visited by boaters during the summer months. Monitoring of at-risk species should continue.

Fire Island (East of Lighthouse)
Brookhaven and Islip, Suffolk County

5,370 acres
0-40' elevation

40.6914°N
72.9846°W

IBA Criteria Met

Criterion	*Species*	*Data*	*Season*	*Source*
Species at Risk	Piping Plover	2 pairs in 1999, 3 in 1998, 5 in 1997, 7 in 1996, 13 in 1995, 7 in 1994, 9 in 1993	Breeding	NY Natural Heritage Biodiversity Databases
Species at Risk	Common Tern	1,221 pairs in 1999, 114 in 1998, 7 in 1997, 258 in 1996, 9 in 1995, 51 in 1994, 181 in 1993	Breeding	NY Natural Heritage Biodiversity Databases
Species at Risk	Least Tern	28 pairs in 1999, 30 in 1998, 52 in 1997, 72 in 1996, 31 in 1995, 22 in 1994, 90 in 1993	Breeding	NY Natural Heritage Biodiversity Databases
Responsibility Species Assemblage-Beach/Dune	Piping Plover, American Oystercatcher, Common Tern, Least Tern, Black Skimmer	Breed	Breeding	NY BBA 2000
Congregations-Waterbirds	Roseate, Common, and Least Terns	1,256 pairs in 1999, 144 in 1998, 59 in 1997, 330 in 1996, 40 in 1995, 73 in 1994, 271 in 1993	Breeding	NY Natural Heritage Biodiversity Databases
Congregations-Individual Species	Common Tern	Supports over 1% (250 pairs) of state estimated breeding population	Breeding	NY Natural Heritage Biodiversity Databases
Congregations-Individual Species	Least Tern	Supports over 1% (25 pairs) of state estimated breeding population	Breeding	NY Natural Heritage Biodiversity Databases

© Jerry and Sherry Liguori

American Oystercatcher

Description: This site includes all but the westernmost few miles of Fire Island, a 32-mile long, quarter-mile wide barrier beach island off the southern shore of Long Island. According to the NY GAP land cover data, over 15% of this site is beach/dune habitat. A site almost eight miles long–the Otis G. Pike Wilderness Area—is the only federal wilderness area in the state. A number of small communities are scattered along the island. The Fire Island Lighthouse, located five miles from the western tip, is located near a bird banding station and hawk watch site.

Birds: This site supports colonial nesting species, including Piping Plovers, Common Terns, and Least Terns. The site serves as a raptor migration corridor, with an average of 5,000 hawks and a maximum of 6,654 between 1980 and 1995. Especially high numbers of American Kestrels (average 2,400; maximum 3,523), Merlins (average 1,230; maximum 1,638), and Peregrine Falcons (average 146; maximum 249) have been documented. The area is a stopover for diverse passerine migrants, with thousands of birds visiting in the fall. A full-scale banding operation that had ceased for several years recently resumed.

Conservation: This site is listed in the 2002 Open Space Conservation Plan as a priority site under the project name Long Island South Shore Estuary Preserve. The primary threat to this area is residential development, with accompanying impacts to local groundwater and loss of habitat. The National Seashore receives about 500,000 visitors per year. Disturbance of nesting terns and plovers by visitors is a potential problem that should be monitored. Hawk watching and banding operations should continue. The extensively planted black pines that provide food and shelter for migrant songbirds along the barrier beach islands are dying and should be replaced with native, salt-tolerant species. During the first round of IBA site identifications, this site was recognized under the research criterion because a long-term monitoring project is based there.

Gardiner's Island

East Hampton, Suffolk County

4,200 acres
0-100' elevation

41.0910°N
72.0958°W

IBA Criteria Met

Criterion	*Species*	*Data*	*Season*	*Source*
Species at Risk	Osprey	34 pairs in 1983	Breeding	Nomination, Steven Biasetti
Species at Risk	Roseate Tern	3 pairs in 1999, 33 in 1997, 5 in 1996, 100 in 1995, 14 in 1994	Breeding	NY Natural Heritage Biodiversity Databases
Species at Risk	Common Tern	70 pairs in 1999, 245 in 1997, 284 in 1996, 73 in 1995, 90 in 1994	Breeding	NY Natural Heritage Biodiversity Databases
Species at Risk	Least Tern	39 pairs in 1999, 69 in 1997, 49 in 1996, 158 in 1995, 180 in 1994	Breeding	NY Natural Heritage Biodiversity Databases
Congregations-Waterfowl	Double-crested Cormorant	Estimated 1,500 pairs in 1993, 1,000 in 1995, 1,545 in 1998	Breeding	Long Island Colonial Waterbird and Piping Plover Survey
Congregations-Waterbirds	Roseate, Common, and Least Terns	284 breeding pairs in 1994, 331 in 1995, 338 in 1996, 347 in 1997, 212 in 1999	Breeding	Long Island Colonial Waterbird and Piping Plover Survey
Congregations-Wading Birds	Great Egret, Snowy Egret, Black-crowned Night-Heron, Glossy Ibis	150 adults in 1993, 245 pairs in 1995, 47 pairs in 1998	Breeding	Long Island Colonial Waterbird and Piping Plover Survey
Congregations-Individual Species	Double-crested Cormorant	Supported approx. 50% of Long Island population in 1993, 30% 1995, over 50% 1998	Breeding	Long Island Colonial Waterbird and Piping Plover Survey
Congregations-Individual Species	Great Egret	Supported over 1% of Long Island population in 1993, 1995, 1998	Breeding	Long Island Colonial Waterbird and Piping Plover Survey
Congregations-Individual Species	Great Black-backed Gull	Supported over 50% of Long Island population in 1998 and 1995	Breeding	Long Island Colonial Waterbird and Piping Plover Survey

Description: This site is a large, privately-owned island located between the tips of the North and South Forks of Long Island, between Gardiners Bay to the west and the open waters of Block Island Sound to the east. The island has a variety of habitats, including ocean bluffs, salt marshes, deciduous forests, brackish ponds, ocean beaches, and maritime grasslands.

Birds: Significant numbers of wintering waterfowl occur here. For example, 1,045 American Black Ducks were counted here in December 1995 as part of the Montauk Christmas Bird Count. The island contains one of the largest concentrations of nesting Ospreys in the state and possibly the largest on the east coast, with 34 active nests in 1983. Historically, over 300 pairs of Ospreys nested on the island. Colonial nesting birds breeding here include the Double-crested Cormorant (1,545 pairs in 1998; over 50% of Long Island's population), Great Egret (30 pairs in 1998), Snowy Egret (10 pairs in 1998), Black-crowned Night-Heron (5 pairs in 1998), Glossy Ibis (two pairs in 1998), American Oystercatcher (eight pairs in 1995), Herring Gull (1,057 pairs in 1998), Great Black-backed Gull (2,000 pairs in 1998), Roseate Tern (three pairs in 1999), Common Tern (70 pairs in 1999), Least Tern (39 pairs in 1999), and Black Skimmer (13 pairs in 1996). The Northern Harrier and Grasshopper Sparrow are confirmed breeders, and Barn Owl, Horned Lark, Yellow-breasted Chat, and Seaside Sparrow are probable breeders.

Conservation: This site is listed in the 2002 Open Space Conservation Plan as a priority site under the project name Peconic Pinelands Maritime Reserve Project. One family has owned the island for over three centuries and natural areas on the island are largely undisturbed. Ideas for increasing recreational use or development should be evaluated with regard to how they impact birds and bird conservation.

Great Gull Island

Southold, Suffolk County

3,700 acres
0-25' elevation

41.2209°N
72.1122°W

IBA Criteria Met

Criterion	*Species*	*Data*	*Season*	*Source*
Species at Risk	Roseate Tern	Estimated 1,747 pairs in 1999, 1,690 in 1998, 1,455 in 1997, 1,064 in 1996, 1,056 in 1995, 1,138 in 1994, 1,400 in 1993	Breeding	Long Island Colonial Waterbird and Piping Plover Survey
Species at Risk	Common Tern	Estimated 10,000 pairs in 1999, 10,000 in 1998, 11,248 in 1997, 9,000 in 1996, 8,000 in 1995, 7,750 in 1994, 7,800 in 1993	Breeding	Long Island Colonial Waterbird and Piping Plover Survey
Congregations-Waterbirds	Mixed species	Has well exceeded threshold (100 ind.) over past ten years	Breeding	Long Island Colonial Waterbird and Piping Plover Survey
Congregations-Individual Species	Roseate Tern	Has supported 60-90% of Long Island population over the past ten years	Breeding	Long Island Colonial Waterbird and Piping Plover Survey
Congregations-Individual Species	Common Tern	Has supported 30-50% of Long Island population over the past ten years	Breeding	Long Island Colonial Waterbird and Piping Plover Survey

Description: This site includes a seven-acre rocky island covered with grassy and herbaceous vegetation (owned by the American Museum of Natural History), and the surrounding marine waters, including a deepwater channel known as "The Race." The Race hosts large concentrations of striped bass (*Morone saxatilis*), bluefish (*Pomatomus saltatrix*), tautog (*Tautoga onitis*), and summer flounder (*Paralichthys dentatus*). It is also a major migration corridor for striped bass and supports a regionally significant commercial lobster (*Homarus americanus*) fishery.

Birds: This is one of the most important tern nesting sites in the world, with the largest breeding colony of Roseate Terns in North America (1,500 pairs in 1996; 45% of the northeast North American population) and one of the largest colonies of Common Terns as well (8,000 pairs in 1995, 7,750 in 1996; 40-45% of the state population). The numbers of terns during the years 2000-2004 continued to be about 10,000 Common Tern nests and roughly 1,600 Roseate Tern nests annually.

Conservation: This site is listed in the 2002 Open Space Conservation Plan as a priority site under the project name Long Island Sound Coastal Area. Oil spills are a potential threat to this area. A long-term research project managed by staff from the American Museum of Natural History involves the demography and life history of terns. During the first round of IBA site identifications, this site was recognized under the research criterion because of the research being performed there.

Great South Bay

Multiple municipalities, Suffolk County

57,000 acres
0-15' elevation

40.6881°N
73.1099°W

IBA Criteria Met

Criterion	*Species*	*Data*	*Season*	*Source*
Species at Risk	Brant	220 ind. in 1999, 215 in 2000	Winter	NYSOA winter waterfowl counts
Species at Risk	American Black Duck	1,095 ind. in 1999, 972 in 2000, 31 in 2001	Winter	NYSOA winter waterfowl counts
Species at Risk	Least Tern	6 pairs in 1997, 15 in 1996, 7 in 1995, 18 in 1994, 17 in 1993	Breeding	NY Natural Heritage Biodiversity Databases
Congregations-Waterfowl	Mixed species	5,681 ind. in 2004, 8,296 in 2003, 8,707 in 2002, 1,652 in 2001, 9,019 in 2000, 3,685 in 1999	Winter	NYSOA winter waterfowl counts
Congregations-Shorebirds	Mixed species	196 ind. on 5/18/95, 528 on 8/18/94, 408 on 7/24/93, 617 on 8/10/92, 1,416 on 8/11/91	Migration	International Shorebird Surveys (ISS)

Description: This site is a protected, open water bay behind Fire Island and Jones Beach Islands, extending roughly from the Nassau/Suffolk County line in the west to Bellport Bay in the east, including eastern Jones Beach (Gilgo and Cedar Beaches). It is the largest shallow saltwater bay in the state, with sandy shoals and extensive eelgrass beds. Great South Bay is a highly productive ecosystem and supports a regionally important commercial and recreational fishery. Sea turtles, including the Atlantic Ridley's turtle (*Lepidochelys kempi*), loggerhead turtle (*Caretta caretta*) and green turtle (*Chelonia mydas*), regularly forage in the area.

Birds: This is a very important waterfowl wintering area. It supports an estimated 25% of the state's wintering American Black Ducks and 22% of the state's wintering scaup, according to an analysis done by the NYS DEC using aerial waterfowl surveys from 1973-1994. The Captree Christmas Bird Count, which covers a portion of the site, has documented averages from 1980-1989 of 1,842 (maximum 3,379) Brants; 1,501 (maximum 2,383) American Black Ducks; and 8,262 (maximum 18,028) Greater Scaup.

Conservation: This site is listed in the 2002 Open Space Conservation Plan as a priority site under the project name Long Island South Shore Estuary Preserve. Portions of this site have been designated as a state Bird Conservation Area. Nutrients input to Great South Bay from point and non-point sources need to be greatly reduced. Degradation of water quality, especially by non-point source runoff, is of mounting concern. Great South Bay is the receptacle for water from more than one million people who live within the bay's drainage basin. Non-point sources dominate the releases into the bay, producing nutrient loading that is followed by eutrophication and increased levels of fecal bacteria. Periodic noxious phytoplankton blooms (brown tides) have major impacts on scallops and other invertebrates, fish, and other wildlife. The cause of these blooms has not yet been established. Public outreach efforts should be focused on reducing fertilizer and pesticide use and proper maintenance of septic systems by landowners in the surrounding upland subdivisions. Monitoring of winter waterfowl should continue.

Harbor Herons Complex
Staten Island, Richmond County

565 acres
0-25' elevation

40.6119°N
74.2006°W

IBA Criteria Met

Criterion	*Species*	*Data*	*Season*	*Source*
Congregations-Wading Birds	Mixed species	Estimated 15-30 in 2004, not surveyed in 2003, 0 nests in 2002, 260 in 2001, 596 in 2000, 763 in 1999 [1]; estimated pairs: 382 in 1998, not surveyed 1996-97, 597 in 1995, 624 in 1994, 730 in 1993, 736 in 1992 [2]	Breeding	1. NY City Audubon Harbor Herons report; 2. NY Natural Heritage Biodiversity Databases
Congregations-Individual Species	Double-crested Cormorant	Supported approx. 40% of Long Island population in 1998, 30% in 1995, 40% in 1993	Breeding	NY City Audubon Harbor Herons report and Long Island Colonial Waterbird and Piping Plover Survey
Congregations-Individual Species	Snowy Egret	Supported approx. 25% of state population in 1998, 25% in 1995, 35% in 1993	Breeding	NY City Audubon Harbor Herons report and Long Island Colonial Waterbird and Piping Plover Survey
Congregations-Individual Species	Cattle Egret	Supported approx. 45% of Long Island population in 1998, 60% in 1995, 95% in 1993	Breeding	NY City Audubon Harbor Herons report and Long Island Colonial Waterbird and Piping Plover Survey
Congregations-Individual Species	Glossy Ibis	Supported approx. 35% of Long Island population in 1998, 35% in 1995, 60% in 1993	Breeding	NY City Audubon Harbor Herons report and Long Island Colonial Waterbird and Piping Plover Survey

Description: This site includes three isolated islands (Isle of Meadows, Shooters Island, and Pralls Island) off the northwestern shore of Staten Island, all owned by the New York City Department of Parks and Recreation. Historically it has been a breeding site for nine species of colonial waterbirds and supported large proportions of the state's wading birds.

Birds: The islands and associated wetlands found here have supported large proportions of the state's wading birds. NY City Audubon and the NYS DEC Long Island Colonial Waterbird and Piping Plover (LICWPP) 1995 survey documented that this area supported 21% of the state's Great Egrets (108 pairs), 57% of the state's Cattle Egrets (25 pairs), 28% of the state's Snowy Egrets (192 pairs), 57% of the state's Black-crowned Night-Herons (643 pairs), 35% of the state's Glossy Ibis (263 pairs), as well as smaller numbers of Double-crested Cormorants (121 pairs), Little Blue Herons (two pairs), Yellow-crowned Night-Herons (14 pairs), Herring Gulls (619 pairs), and Great Black-backed Gulls (10 plus pairs). In recent years, Shooters Island and Isle of Meadows no longer support breeding wading birds. In 2004, Pralls Island supported 15 Black-crowned Night-Herons, although nesting did not appear to be successful. In 2004, there was a breeding pair of Osprey and there were 31 active Double-crested Cormorant nests on Shooters Island.

Conservation: This site is listed in the 2002 Open Space Conservation Plan as a priority site under the project name Harbor Herons Wildlife Complex. Toxicology studies have found high levels of heavy metals and DDT in unhatched Glossy Ibis eggs, a condition that reduces successful hatching rates. Success of the colonies requires ample foraging areas, many of which remain unprotected. Protection of these wetland foraging sites and clean-up of contaminants in other foraging areas should be a high priority. New York City Audubon has surveyed this site as part of their Harbor Herons surveys, and monitoring should continue. During the first round of IBA site identifications, this site was recognized under the research criterion because a long-term monitoring project is based there

Hempstead Lake State Park

Hempstead, Nassau County

500 acres
15-50' elevation

40.6833°N
73.6416°W

IBA Criteria Met

Criterion	Species	Data	Season	Source
Congregations-Waterfowl	Mixed species	Peak numbers of waterfowl in late fall and winter reach many thousands [1]. Thousands of American Black Ducks winter on the ponds with huge numbers of other ducks, and hundreds of Canada Geese [2].	Migration and winter	1. Nomination, Emanuel Levine; 2. Sy Schiff pers. comm. 2004
Congregations-Migrant Landbirds	Mixed species	Large numbers of flycatchers, thrushes, warblers, tanagers and a large mix of other birds are found daily from mid-April to late May.	Migration	Sy Schiff pers. comm. 2004

Description: This site is administered by NYS OPRHP and includes a one-mile-long, half-mile-wide lake and surrounding lands. The lake provides a variety of aquatic habitats and is used by wintering waterfowl. The park's additional bodies of water, South and Sailboat Ponds, are less attractive for waterfowl and receive less use.

Birds: Hempstead Lake is one of the most important sites on Long Island for wintering waterfowl, with buildups beginning in late August and peaking in the late fall and winter. At peak times, the numbers run into the many thousands with the following species present: Gadwall, American Wigeon, American Black Duck, Mallard, Northern Shoveler, Northern Pintail, Green-winged Teal, Canvasback, Lesser Scaup, Common Merganser, Hooded Merganser, and Ruddy Duck. Of these, the most numerous are the American Black Duck, Mallard, and Lesser Scaup. While the American Black Ducks and Mallards move in and out all day, the Lesser Scaup (which in some years have totaled several thousand) stay on the lake continuously. This is also one of the most important sites for migrant landbirds on Long Island. A normal day reveals 50-75 species of birds during a leisurely morning stroll. In addition, approximately 17 species of shorebirds have been observed foraging at the north end of the lake when water levels go

American Black Duck

down. Large numbers of Common Terns and some Forster's Terns use the area as a feeding and bathing site in late summer.

Conservation: A proposal has been made to open the lake to boating for fishing purposes. This may cause disturbance to waterfowl if permitted during the winter season. A thorough assessment of the potential impacts of boating on wintering waterfowl should be conducted before the activity is approved. Depending on the findings of such an assessment, it may be necessary to control boating during the peak waterfowl season. Regular surveys of waterfowl should be carried out at the site.

Hoffman and Swinburne Island Complex

Staten Island, Richmond County

11 acres
10' elevation

40.5725°N
74.0527°W

IBA Criteria Met

Criterion	*Species*	*Data*	*Season*	*Source*
Congregations-Wading Birds	Mixed species	500 nests in 2004, 472 in 2003, 513 in 2002, 403 in 2001 [1]; 20 estimated nesting pairs in 1998, not surveyed 1996-97, 12 in 1995, not surveyed 1994, 6 in 1993, 1 in 1991 [2]	Breeding	1. NY City Audubon Harbor Herons report; 2. NY Natural Heritage Biodiversity Databases

Snowy Egret

Description: This site includes two artificial islands, Hoffman and Swinburne, in the lower New York Bay, which are owned by the National Park Service as part of Gateway National Recreation Area. The complex is undergoing rapid succession to deciduous forest, with tall shrubs encroaching on the few remaining open grassy areas. The habitat consists primarily of thick vegetation surrounded by rocky shoreline. This extremely suitable habitat has seen an explosion in its wading bird population in the last three years (2000-2003). The only structures are two brick buildings on Swinburne and a few piers on the western shores of both islands—all in disrepair.

Birds: Hoffman Island was the second largest heronry in the New York/New Jersey harbor in 2004 supporting six species of long-legged waders: the Great Egret, Snowy Egret, Little Blue Heron, Cattle Egret, Black-crowned Night-Heron, Yellow-crowned Night-Heron, and Glossy Ibis. Herring and Great Black-backed Gulls also nest on both islands; in 2004, the islands supported 131 Herring Gulls and 118 Great Black-backed Gulls. Hoffman and Swinburne Islands also support nesting Double-crested Cormorants. The population on Hoffman Island increased in 2004 to 34 nests and the Swinburne population appeared to decline to 108 nests. In recent years, it has been the only island supporting a nesting population of Cattle Egrets. The islands' fruit-bearing trees and plants attract migrant birds that likely use the islands as a stopover point.

Conservation: Suggestions have been made to remove the invasive ailanthus (*Ailanthus altissima*) and black locust (*Robinia pseudoacacia*) trees, and to plant gray birch and black cherry trees to provide good nesting areas. The islands are undisturbed with no plans for development. Bird surveys of the Hoffman & Swinburne Island Complex by New York City Audubon are conducted yearly.

Huntington and Northport Bays

Huntington, Suffolk County

13,400 acres
0-145' elevation

40.9289°N
73.4253°W

IBA Criteria Met

Criterion	*Species*	*Data*	*Season*	*Source*
Species at Risk	American Black Duck	828 ind. in 1999, 1,430 in 2000, 316 in 2001	Winter	NYSOA winter waterfowl counts
Species at Risk	Piping Plover	7 pairs in 2001, 6 in 2000, 15 in 1999, 8 in 1998, 6 in 1997, 10 in 1996, 8 in 1995, 3 in 1994, 3 in 1993	Breeding	NY Natural Heritage Biodiversity Databases
Species at Risk	Common Tern	13 pairs in 2001, 0 in 2000, 810 in 1999, 790 in 1998, 630 in 1997, 540 in 1996, 765 in 1995, 495 in 1994, 135 in 1993	Breeding	NY Natural Heritage Biodiversity Databases
Species at Risk	Least Tern	86 pairs in 2001, 41 in 2000, 56 in 1999, 114 in 1998, 28 in 1997, 157 in 1996, 338 in 1995, 274 in 1994, 137 in 1993	Breeding	NY Natural Heritage Biodiversity Databases
Species at Risk	Black Skimmer	5 pairs in 1999, 11 in 1998, 10 in 1997, 9 in 1996, 7 in 1995, 3 in 1994, 0 in 1993	Breeding	NY Natural Heritage Biodiversity Databases
Congregations-Waterfowl	Mixed Species	2,252 ind. in 2004, 3,847 in 2003, 2,582 in 2002, 5,012 in 2001, 6,114 in 2000, 3,637 in 1999	Winter	NYSOA winter waterfowl counts
Congregations-Waterbirds	Terns	Estimated 99 pairs in 2001, 41 in 2000, 866 in 1999, 904 in 1998, 658 in 1997, 697 in 1996, 1,103 in 1995, 769 in 1994, 272 in 1993	Breeding	NY Natural Heritage Biodiversity Databases

Description: This site covers an area on Long Island Sound that extends from Cold Spring Harbor in the west to Eaton's Neck and Asharoken in the east, and includes offshore waters. The site includes Caumsett State Historic Park (administered by NYS OPRHP), Target Rock National Wildlife Refuge (USFWS), and other municipally owned and privately owned areas. The site includes dunes, sand spits, vegetated spoil islands, tidal basins, saltwater marshes, and open water. Caumsett State Historic Park has large fields and freshwater wetlands. Much of the land outside of the parks and refuges is heavily developed.

Birds: The site is an important waterfowl wintering area, with 6,114 individuals seen on the 2000 New York State Ornithological Association (NYSOA) winter waterfowl count. Colonial nesting birds breeding here include Piping Plovers, Common Terns, Least Terns, and Black Skimmers. Other at-risk species supported at the site include the Common Loon (winters), Osprey (confirmed breeder), American Oystercatcher (confirmed breeder), Willow Flycatcher (probable breeder), Wood Thrush (confirmed breeder), and Blue-winged Warbler (confirmed breeder). Wooded and shrub habitats along the shoreline are likely important migrant landbird stopover areas.

Conservation: This site is listed in the 2002 Open Space Conservation Plan as a priority site under the project name Long Island Sound Coastal Area. Portions of Caumsett State Historic Park have been designated a state Bird Conservation Area. Near Lloyd Point in Caumsett State Historic Park, the tidal basin can attract up to 200 boats. Illegal landings on shoreline areas represent a significant threat to plovers and plover habitat. Zodiacs and jet skis have the potential to disturb nesting birds, specifically Piping Plovers. Unauthorized camping and dogs can disturb nesting birds. Similar problems exist at Hobart Beach Preserve and near the Eaton's Neck Coast Guard Station. Exclosures and fencing have helped, but there is a continuing need for public education about the significance of the area, and patrolling to ensure that nesting birds are not disturbed. The plover stewardship program overseen by NYS OPRHP in cooperation with the Krusos Foundation should continue. At Hobart Beach, vegetation growth may be making the area less suitable for nesting terns. Pollution from sewage, storm water runoff, and other sources has detrimental impacts on aquatic systems, and may impact food supplies and the health of wintering waterfowl. A cooperative effort to clean up Long Island Sound is underway. Remaining open spaces along the shoreline are subject to great pressure for residential development. A program to encourage private landowners to manage habitats for migratory birds would be beneficial. Inventory and monitoring, particularly of at-risk species, should continue.

Jamaica Bay

Brooklyn and Queens, Kings and Queens Counties

21,000 acres
0-15' elevation

40.6080°N
73.8693°W

IBA Criteria Met

Criterion	*Species*	*Data*	*Season*	*Source*
Species at Risk	Brant	433 ind. in 2000, 290 in 2001	Winter	NYSOA winter waterfowl counts
Species at Risk	American Black Duck	849 ind. in 2000, 88 in 2001	Migration and winter	NYSOA winter waterfowl counts
Species at Risk	Osprey	8 pairs	Breeding	Don Reipe pers. comm. 2004
Species at Risk	Piping Plover	8 pairs in 1999, 11 in 1998, 10 in 1997, 17 in 1996, 23 in 1995, 29 in 1994, 23 in 1993	Breeding	NY Natural Heritage Biodiversity Databases
Species at Risk	American Oystercatcher	100 pairs	Breeding	Don Reipe pers. comm. 2004
Species at Risk	Common Tern	1,620 pairs in 1999, 3,060 in 1998, 1,950 in 1997, 2,028 in 1996, 2,665 in 1995, 3,360 in 1994, 3,160 in 1993	Breeding	NY Natural Heritage Biodiversity Databases
Species at Risk	Black Skimmer	200 pairs	Breeding	Don Reipe pers. comm. 2004
Responsibility Species Assemblage-Wetlands	American Black Duck, Glossy Ibis, Clapper Rail, Marsh Wren, Saltmarsh Sharp-tailed Sparrow, Seaside Sparrow	Breed	Breeding	NY BBA 2000
Congregations-Waterfowl	Mixed species	2,403 ind. in 2003, 8,330 in 2002, 1,027 in 2001, 2,475 in 2000	Winter	NYSOA winter waterfowl counts
Congregations-Waterbirds	Gulls	Estimated 8,585 nesting pairs in 1993, 7,972 in 1994, 8,485 in 1995, 4,826 in 1996, 6,331 in 1997, 5,448 in 1998	Breeding	NY Natural Heritage Biodiversity Databases

Criterion	*Species*	*Data*	*Season*	*Source*
Congregations-Waterbirds	Terns	Estimated 1,620 nesting pairs in 1999 [1]; 936 in 1993, 628 in 1994, 385 in 1995, 361 in 1996 [2]	Breeding	1. NY Natural Heritage Biodiversity Databases; 2. Brown *et al.* 2001
Congregations-Wading Birds	Mixed species	544 nests in 2004, 504 in 2003, 180 in 2002, 156 in 2001	Breeding	NY City Audubon Harbor Herons report (Canarsie Pol)
Congregations-Shorebirds	Mixed species	Over 1,000 shorebirds present on many days during migration	Migration	Don Reipe pers. comm. 2004

Description:

The Jamaica Bay complex includes the marine and tidal wetland portions of the bay itself as well as the barrier beach/dune system and some adjoining upland shrub and grassland. Jamaica Bay is a saline to brackish, eutrophic estuary with a mean depth of 13 feet. According to the NY GAP land cover data, the site includes approximately 2,000 acres of salt marsh habitat. It is situated in the midst of the New York City metropolitan area. The uplands around the bay, as well as much of the Rockaway barrier beach are developed. About 12,000 of the original 16,000 acres of wetlands in the bay have been filled in. Extensive areas have been dredged for navigation channels and to provide fill for airports and other construction projects. The site hosts breeding diamondback terrapins (*Malaclemys terrapin*) and plants such as seabeach amaranth (*Amaranthus pumilus*), seabeach knotweed (*Polygonum glaucum*), Schweinitz's flatsedge (*Cyperus schweinitzii*), and slender flatsedge (*Cyperus filiculmis*). U.S. National Park Service's Gateway Recreation Area encompasses the largest portion of wildlife habitat, but some land is owned or administered by the New York City Department of Parks and Recreation, NYS OPRHP, and many private landowners.

Jamaica Bay

Birds: This is a critical saltwater wetland habitat supporting a renowned abundance and diversity of shorebirds, waterfowl, gulls, terns, and other species. During migration, the site hosts 600-1,200 Black-bellied Plovers (1% or more of the North American population), 200-1,600 Red Knots (1% or more of the eastern flyway population), and more than 35 other shorebird species. The beaches are breeding sites for 20-30 (22 in 1996) pairs of Piping Plovers (1% or more of the east coast population; 9% of the state population); 4,500-6,000 (5,830 in 1995) pairs of Laughing Gulls (99% of state population); 2-4 pairs of Roseate Terns; 2,000-3,000 (2,078 in 1996, 2,737 in 1995) pairs of Common Terns (1% or more of east coast population; 11-16% of state population); 30-80 (38 in 1996, 77 in 1995) pairs of Forster's Terns; 70-200 (73 in 1996, 189 in 1994) pairs of Least Terns (2-7% of state population); and 190-250 (250 in 1996) pairs of Black Skimmers (51% of state population). The area is an important waterfowl wintering area as well, with healthy numbers of Brant and scaup. A hawk watch within the complex tallies 5,000-6,000 plus hawks each year, including several Peregrine Falcons.

Conservation: This site is listed in the 2002 Open Space Conservation Plan as a priority site under the project name Jamaica Bay Protection Area. Development of remaining open space within the complex is a concern. A 302-acre site that hosts breeding Piping Plovers, along the Atlantic Ocean beaches of the Rockaway Peninsula, is being considered for development. Pollution is an ongoing problem, though great improvements have been made in recent decades. Sewage, storm drain outflow, and contaminated sediments are ongoing issues. Swimming and shell fishing are prohibited, and health advisories warn against fish consumption. The portions of the complex owned and managed by the National Park Service as part of the Gateway National Recreation Area are under competing pressures for various public recreational uses, including biking, surf-fishing, sunbathing, swimming, concerts, and educational programs. These uses may sometimes conflict with the needs of birds and other wildlife and should be weighed carefully. Serious concern has been raised about disappearing marsh habitat, and a number of organizations are involved in researching and addressing this issue. A symposium and public forum on the disappearing marsh issue was held in 2004. New York City Audubon has surveyed Canarsie Pol as part of their Harbor Herons surveys and monitoring should continue.

Little Neck Bay to Hempstead Harbor

Multiple municipalities, Queens and Nassau Counties

13,300 acres
0-55' elevation

40.8256°N
73.7128°W

IBA Criteria Met

Criterion	*Species*	*Data*	*Season*	*Source*
Species at Risk	American Black Duck	1,298 ind. in 1996, 591 in 1999, 309 in 2000, 810 in 2001	Winter	NYSOA winter waterfowl counts
Species at Risk	Piping Plover	3 pairs in 2002, 4 in 2001, 0 pairs in 1999, 0 in 1997, 2 ind. in 1993	Breeding	NY Natural Heritage Biodiversity Databases
Species at Risk	Least Tern	135 pairs in 2002, 74 in 2001, 0 in 1999, 16 in 1997, 13 in 1996, 0 in 1995, 0 in 1994, 0 in 1993	Breeding	NY Natural Heritage Biodiversity Databases
Congregations-Waterfowl	Mixed species	2,177 ind. in 2004, 6,039 in 2003, 2,687 in 2002, 3,966 in 2001, 5,267 in 2000, 11,598 in 1999	Winter	NYSOA winter waterfowl counts
Congregations-Individual Species	Greater Scaup	Site has supported over 1% of estimated state wintering population. 502 ind. in 2004, 960 in 2003, 1,150 in 2001, 234 in 2000, 6,183 in 1999	Winter	NYSOA winter waterfowl counts

Description: This site includes the waters and adjoining wetlands on the southernmost corner of Long Island Sound. The site extends from Little Neck Bay east to Hempstead Harbor. Underwater lands are largely under the jurisdiction of the Town of North Hempstead.

Birds: This site is an important waterfowl wintering area. The NYSOA winter waterfowl survey documented 11,598 waterfowl in 1999, which was largely composed of Greater Scaup (6,183), one of the largest concentrations in the state. The site is also an important breeding area for Piping Plovers and Least Terns. Other at-risk species that have been documented here include the Brant (winters), American Black Duck (common in winter), Common Loon (winters), Pied-billed Grebe (winters), American Bittern (rare migrant), Osprey (breeder, more than four nests), Northern Harrier (migrant), Sharp-shinned Hawk (migrant and winters), Northern Goshawk (rare migrant), Peregrine

Falcon (migrant), American Oystercatcher (regular breeder), Common Tern (common migrant), Black Tern (uncommon migrant), Black Skimmer (rare migrant), Common Nighthawk (migrant), Yellow-breasted Chat (rare migrant), Vesper Sparrow (rare migrant), Nelson's Sharp-tailed Sparrow (migrant), Seaside Sparrow (rare migrant), plus other rarities. Over the past 30 years, over 250 species have been documented at Sands Point, which comprises a variety of habitats unique in Nassau County. The site is also an important stopover site for shorebirds; 100-200 individuals are regularly seen at Prospect Point alone during fall migration.

Conservation: This site is listed in the 2002 Open Space Conservation Plan as a priority site under the project name Long Island Sound Coastal Area. Water pollution from various sources, including contaminants, oil spills, suburban runoff, excessive sedimentation, and sewage and storm water discharges, is a major factor affecting aquatic resources on which waterfowl and other waterbirds rely. Increased residential and commercial development along the shorelines may cause increased pollution and disturbance of birds. Protection of significant sites, particularly wetlands, should continue. Smaller wetlands are susceptible to development and need better protection. Waterfowl numbers should be monitored regularly. Existing monitoring includes Piping Plover and Least Tern surveys at Prospect Point, Plum Point, and Half Moon Beach in Sands Point.

Long Island Pine Barrens
Multiple municipalities, Suffolk County

96,000 acres
0-245' elevation

40.8801°N
72.7918°W

IBA Criteria Met

Criterion	*Species*	*Data*	*Season*	*Source*
Species at Risk	Whip-poor-will	Extent of habitat and breeding atlas presence strongly suggest that the threshold is being met [1]: 8PR, 2PO [2]; 1CO, 10PR [3]	Breeding	1. Technical Committee consensus; 2. NY BBA 2000; 3. NY BBA 1980
Species at Risk	Blue-winged Warbler	3CO, 7PR [1]; 11CO [2]	Breeding	NY BBA 2000; NY BBA 1980
Species at Risk	Grasshopper Sparrow	Extent of habitat and breeding atlas presence strongly suggest that the threshold is being met [1]; 225 ind. in 2000	Breeding	1. Technical Committee consensus; 2. Pelkowski 2001
Responsibility Species Assemblage-Shrub	Northern Bobwhite, American Woodcock, Whip-poor-will, Eastern Kingbird, Gray Catbird, Brown Thrasher, Blue-winged Warbler, Prairie Warbler, Eastern Towhee, Field Sparrow	Breed	Breeding	NY BBA 2000
Responsibility Species Assemblage-Forest	Broad-winged Hawk, Black-billed Cuckoo, Hairy Woodpecker, Northern Flicker, Eastern Wood-Pewee, Great Crested Flycatcher, Yellow-throated Vireo, Wood Thrush, Black-and-white Warbler, Scarlet Tanager, Rose-breasted Grosbeak, Baltimore Oriole	Breed	Breeding	NY BBA 2000

Description: This site includes a fairly contiguous expanse of pitch pine/oak forest with coastal ponds, red maple swamps, and other wetlands extending from central Long Island to the South Fork, and varying from 2.5 to 9 miles wide. According to the NY GAP land cover data, approximately

65% of this site is pitch pine oak, which supports both forest and shrub species. Other forest habitat includes oak and successional hardwoods. Other shrub habitat includes successional shrub, shrub swamp, old/field pasture, and cropland. The area supports one of the highest diversities of rare plant and animal species in the state. It is owned by a mix of private, municipal, state, federal, non-governmental conservation agencies.

Birds: This is the largest pine barren ecosystem in New York, supporting at-risk species (including breeding populations) such as the American Black Duck (confirmed), Osprey (confirmed), Northern Harrier (probable breeder), Sharp-shinned Hawk (possible breeder), Cooper's Hawk (confirmed), Upland Sandpiper (very few in recent years), American Woodcock (confirmed), Common Nighthawk (confirmed), Whip-poor-will (confirmed), Willow Flycatcher (probable breeder), Horned Lark (confirmed), Wood Thrush (confirmed), Blue-winged Warbler (confirmed), Prairie Warbler (confirmed), Worm-eating Warbler (possible breeder), Vesper Sparrow (confirmed), and Grasshopper Sparrow (225 ind. in 2000). The area includes the last remaining viable grassland bird community on Long Island, with breeding Upland Sandpipers, Vesper Sparrows, and Grasshopper Sparrows. Other characteristic pine barren species that are abundant here include Brown Thrashers, Blue-winged Warblers, Pine Warblers, Prairie Warblers, and Field Sparrows.

Conservation: This site is listed in the 2002 Open Space Conservation Plan as a priority site under the project name Pine Barrens Core, CGA & CRA. Development is no longer a major threat because the area was designated a New York State Forest Preserve by executive order in 1995; since then, $32 million has been appropriated for land acquisition. The David A. Sarnoff Pine Barrens Preserve has been designated a state Bird Conservation Area. The Protected Lands Council, a collaboration of land managers and interested citizens, meets regularly to coordinate stewardship of protected lands. Any new or expanded recreation proposals must consider potential impacts on birds and bird habitats. If potential adverse impacts occur, alternative locations or designs must be considered. Long-term threats are over-extraction or pollution of groundwater, and succession caused by fire suppression. The Long Island Pine Barrens has been selected by the Nature Conservancy (TNC) Fire Learning Network as one of five demonstration sites nationwide. TNC has also begun a pilot restoration project with volunteer partners from local schools, businesses, and land management agencies to restore the dwarf pine community.

Mecox Sagaponack Coastal Dunes

Easthampton and Southampton, Suffolk County

3,500 acres
0-30' elevation

40.9001°N
72.3064°W

IBA Criteria Met

Criterion	*Species*	*Data*	*Season*	*Source*
Species at Risk	Piping Plover	1 pair in 2002, 5 in 1999, 7 in 1998, 8 in 1997, 10 in 1996, 5 in 1995, 8 in 1994, 4 in 1993	Breeding	NY Natural Heritage Biodiversity Databases
Species at Risk	Least Tern	15 pairs in 1999, 18 in 1998, 32 in 1997, 64 in 1996, 129 in 1995, 133 in 1994, 72 in 1993	Breeding	NY Natural Heritage Biodiversity Databases

Description: This site includes the coastal beaches and wetlands extending from Watermill Beach in the west to Georgica Pond in the east. The site includes undeveloped flats, sand bars, and an ocean inlet.

Birds: The area is important to breeding Piping Plovers and Least Terns, migrating shorebirds, and wintering waterfowl.

Conservation: Mecox Bay has been designated a significant coastal fish and wildlife habitat by the New York State Department of State due to species vulnerability, ecosystem rarity, and the site's significance as a winter waterfowl area. Nesting populations of Least Terns and Piping Plovers are vulnerable during the nesting season. Therefore, recreational activities including beach-walking, boat landing, off-road vehicle use, and picnicking near nesting areas should be minimized from April through mid-August. Predation by foxes, gulls, and crows is also of concern; flooding and subsequent loss and abandonment of nests can occur at spring high tides.

Montauk Point

East Hampton, Suffolk County

7,800 acres
0-100' elevation

41.0668°N
71.8816°W

IBA Criteria Met

Criterion	*Species*	*Data*	*Season*	*Source*
Species at Risk	Common Loon	314 ind. in 2001	Winter	NYSOA winter waterfowl counts
Responsibility Species Assemblage-Shrub	Northern Bobwhite, American Woodcock, Eastern Kingbird, Gray Catbird, Brown Thrasher, Blue-winged Warbler, Prairie Warbler, Eastern Towhee, Field Sparrow	Breed	Breeding	NY BBA 2000
Congregations-Waterfowl	Mixed species	25,398 ind. in 2002, 29,486 in 2001, 23,440 in 1997	Winter	NYSOA winter waterfowl counts
Congregations-Waterbirds	Pelagic seabirds	Sizable concentrations of pelagic seabirds occur in the waters off the point. 250 ind. Northern Gannets were counted on the Dec. 1995 CBC	Winter	Nomination, Steven Biasetti
Congregations-Individual Species	Surf, White-winged, and Black Scoters	Supported 27% of estimated northeast scoter population in 2002, 57% in 2001, 1% in 1997	Winter	NYSOA winter waterfowl counts and United States Fish and Wildlife Service 2004

Description: This site includes the easternmost point of land on Long Island, extending from Lake Montauk in the west to Montauk Point State Park and including the offshore waters. A large portion of the area is under public ownership, including Montauk Point State Park and Camp Hero State Park. The site contains an impressive diversity of maritime upland, wetland, and shoreline habitats. According to the NY GAP land cover data, over 35% of this site is shrub habitat, which includes pitch pine oak, shrub swamp, and successional hardwoods. The waters off of the point contain extensive mussel and kelp beds and are an important feeding area for juvenile Atlantic ridley turtles (*Lepidochelys kempii*),

Montauk Point

loggerhead turtles (*Caretta caretta*), and leatherback turtles (*Dermochelys coriacea*). Marine mammals including gray seals (*Halichoerus grypus*), harbor seals (*Phoca vitulina*), northern right whales (*Eubalaena glacialis*), finback whales (*Balaenoptera physalus*), humpback whales (*Megaptera novaeangliae*), and minke whales (*Balaenoptera acutorostrata*) regularly forage in or migrate through the near-shore waters

Birds: The point is a very important waterfowl wintering area, with the largest winter concentration of sea ducks in the state. A waterfowl count in January 1997 documented 17,514 Common Eiders, 120 Long-tailed Ducks, 1,900 Surf Scoters, 2,402 White-winged Scoters, 1,000 Black Scoters, and 320 Red-breasted Mergansers. The 1996 NYS DEC mid-winter aerial waterfowl survey documented 4,300 scoters and 250 Long-tailed Ducks. The December 1995 Christmas Bird Count (CBC) tallied 1,500 Greater Scaup, 5,000-plus Common Eiders, 500-plus White-winged Scoters, 600-plus Common Goldeneyes, and 600-plus Red-breasted Mergansers. King Eiders and Harlequin Ducks occur here regularly in winter. Montauk is the southernmost wintering area for Common Eiders and Harlequin Ducks on the East Coast. Sizable concentrations of pelagic seabirds occur in the waters off the point. For example, 250 Northern Gannets were counted in the December 1995 CBC. Wetland areas around Big and Little Reed Ponds support confirmed or probable breeding at-risk species, including the American Black Duck, Least Bittern, Northern Harrier, and Red-shouldered Hawk. Upland areas host characteristic shrub breeding species including the Northern Bobwhite, American Woodcock, Eastern Kingbird, Gray Catbird, Brown Thrasher, Blue-winged Warbler, Prairie Warbler, Eastern Towhee, and Field Sparrow.

Conservation: This site is listed in the 2002 Open Space Conservation Plan as a priority site under the project name Peconic Pinelands Maritime Reserve Project. Much of the area is under public ownership, but significant portions are under private ownership. Poorly planned development could destroy or degrade sensitive habitats, and could lead to increases in non-point source water pollution problems. Further attention needs to be given to protect the offshore waters and associated aquatic habitats around Montauk Point. The area is extremely vulnerable to oil or contaminants spills, which could have devastating impacts on waterfowl and seabirds. Procedures for rapid response and containment in the event of such a spill should be developed and equipment should be readied. The extensively planted black pines are dying and should be replaced with native, salt-tolerant species that provide food and shelter for migrant songbirds. Standardized annual monitoring of migrant and wintering waterfowl and seabirds is encouraged.

Harlequin Ducks

Moriches Bay

Multiple municipalities, Suffolk County

13,000 acres
0-30' elevation

40.7879°N
72.6955°W

IBA Criteria Met

Criterion	*Species*	*Data*	*Season*	*Source*
Species at Risk	Piping Plover	48 pairs in 1998, 42 in 1997, 40 in 1996, 34 in 1995, 25 in 1994, 30 in 1993	Breeding	NY Natural Heritage Biodiversity Databases
Species at Risk	Common Tern	631 pairs in 1999, 1,106 in 1998, 906 in 1997, 1,468 in 1996, 52 in 1995, 66 in 1994, 879 in 1993	Breeding	NY Natural Heritage Biodiversity Databases
Species at Risk	Least Tern	6 pairs in 1999, 19 in 1998, 14 in 1997, 0 in 1996, 531 in 1995, 47 in 1993	Breeding	NY Natural Heritage Biodiversity Databases
Species at Risk	Black Skimmer	0 pairs in 1999, 23 in 1998, 33 in 1997, 13 in 1996, 6 in 1995, 9 in 1994, 27 in 1993	Breeding	NY Natural Heritage Biodiversity Databases
Congregations-Waterfowl	Mixed species	Over 5,000 ind. waterfowl on average, with 8,382 ind. during the peak year	Winter	NYS DEC mid-winter aerial waterfowl surveys for the period of 1975-1984
Congregations-Waterbirds	Terns	Estimated 637 pairs in 1999, 1,129 in 1998, 920 in 1997, 1,504 in 1996, 586 in 1995, 216 in 1994, 948 in 1993	Breeding	NY Natural Heritage Biodiversity Databases
Congregations-Wading Birds	Mixed species	Islands in the bay easily support more than 100 pairs of nesting herons.	Breeding	Mike Wasilco pers. comm. 2004

Description: This site consists of a bay, marsh, and barrier beach complex (with adjoining uplands) on the south shore of Long Island, extending from the Floyd Estate in Mastic (mainland portion of Fire Island National Seashore) in the west to Westhampton Beach in the east. The site includes Haven's Estate and Cupsogue County Park, both owned by Suffolk County. It is a productive area for marine finfish, shellfish, and other wildlife.

Birds: This site is important for nesting wading birds. West Inlet Island alone supports large numbers of Great Egrets (108 pairs in 2004), Snowy Egrets (59 pairs in 2004), Little Blue Heron (two pairs in 2004), Tricolored Heron (one pair in 2004), Black-crowned Night-Heron (155 pairs in 2004), and Glossy Ibis (44 pairs in 2004). The site also supports at risk species such as Osprey (breeds), Piping Plovers (48 pairs in 1998), Roseate Terns (four pairs in 1998), Common Terns (631 pairs in 1999), Least Terns (six pairs in 1999), Black Skimmers (23 pairs in 1998), and Seaside Sparrow (breeds). Herring Gulls (368 pairs in 1995, 3% of the state population.) and Great Black-backed Gulls (168 pairs in 1995; 3% of the state population) nest here as well. The salt marshes support breeding Clapper Rails, American Oystercatchers, Willets, and Saltmarsh Sharp-tailed Sparrows. The site is also an important waterfowl wintering area. NYS DEC mid-winter aerial waterfowl surveys from 1975-1984 documented over 5,000 individuals on average (8,382 in peak year). These included an average of 350 Brant (580 maximum), 400 Canada Geese (870 maximum), 1,100 American Black Ducks (1,580 maximum), 225 Mallards (430 maximum), 2,150 scaup (4,470 maximum), and 400 Red-breasted Mergansers (920 maximum).

Conservation: This site is listed in the 2002 Open Space Conservation Plan as a priority site under the project name Long Island South Shore Estuary Preserve. Portions of this site have been designated as a state Bird Conservation Area. Water pollution, including that caused by chemical contamination, oil spills, and sewage or storm water runoff, will adversely affect the area's biological productivity. Efforts should be made to control waste discharged from recreational boats and upland sources. Construction and maintenance of shoreline structures, including docks, piers, and bulkheads, should be avoided in undisturbed areas. Elimination of salt marsh and intertidal areas through excavation or filling that would decrease available habitat for salt marsh breeding birds should be prevented. As nesting Piping Plovers and Least Terns are highly susceptible to human disturbance, recreational activities, including boat landing, off-road vehicle use, and picnicking, should be minimized near nesting areas from April through mid-August. Fencing, beach closures, posting, beach warden patrols, and public education may be required in some areas. Monitoring of at-risk species and waterfowl should continue.

Muttontown Preserve

Oyster Bay, Nassau County

550 acres
195-245' elevation

40.8269°N
73.5365°W

IBA Criteria Met

Criterion	*Species*	*Data*	*Season*	*Source*
Responsibility Species Assemblage-Shrub	American Woodcock, Eastern Kingbird, Gray Catbird, Brown Thrasher, Blue-winged Warbler, Eastern Towhee, Field Sparrow	Breed	Breeding	Ken Feustel pers. comm. 2004

Eastern Kingbird

Description: This site consists of moist woodlands, fields, pioneer woodlands, conifer stands, kettle ponds, and upland forests. It is owned by the Nassau County Department of Recreation and Parks. The site is a relatively large, contiguous tract of habitat serving as an island in the midst of ongoing development. It contains an area known as the Christie Estate, which includes grassland habitat. According to the NY GAP land cover data, over 50% of this site is shrub habitat, which includes pitch pine oak, successional shrub, and old/field pasture.

Birds: This site offers diversified habitat for a variety of breeding birds. Its importance stems not from rare or endangered species, but from the variety of characteristic species, including the American Woodcock, Yellow-billed Cuckoo, Eastern Wood-Pewee, Eastern Kingbird, White-eyed Vireo, Red-eyed Vireo, Gray Catbird, Brown Thrasher, Blue-winged Warbler, Chestnut-sided Warbler, Black-and-white Warbler, Ovenbird, Rufous-sided Towhee, Field Sparrow, Rose-breasted Grosbeak, Indigo Bunting, and Baltimore Oriole. In addition, it is one of only a few sites in Nassau County where both Great Horned Owls and Eastern Screech-Owls nest.

Conservation: This site is listed in the 2002 Open Space Conservation Plan as a priority site under the project name Long Island Sound Coastal Area. The preserve currently has little funding and there are no plans for activities that would negatively impact the avifauna. Consideration should be given to management of early successional habitats and invasive, non-native plant species. Because of the size of the site and its location in a major urban area, there is potential for the site to be considered for future development of recreational facilities, which might increase human disturbance and decrease available habitat. Monitoring of shrub and at-risk species is needed to better understand bird use and determine if this site continues to meet IBA criteria.

Napeague Harbor and Beach

East Hampton, Suffolk County

4,350 acres
0-20' elevation

40.9958°N
72.0717°W

IBA Criteria Met

Criterion	*Species*	*Data*	*Season*	*Source*
Species at Risk	Piping Plover	1 pair in 2002, 6 in 1999, 11 in 1998, 9 in 1997, 8 in 1996, 12 in 1995, 7 in 1994, 5 in 1993	Breeding	NY Natural Heritage Biodiversity Databases

Description: This site includes the Napeague State Park, administered by NYS OPRHP, and surrounding wetlands and beaches, including Napeague Harbor.

Birds: This site provides important habitat for the Northern Harrier (male and female have been observed), Piping Plover (six pairs in 1999), Common Tern (two pairs in 1997), and Least Tern (five pairs in 1999).

Conservation: Nesting Piping Plovers and terns are vulnerable to disturbance during the nesting season. Therefore, recreational activities including beach-walking, boat landing, off-road vehicle use, and picnicking near nesting areas should be minimized from April through mid-August. Predation by foxes, gulls, and crows is also a concern. Flooding and subsequent loss and abandonment of nests can occur at spring high tides.

Dunes within the Napeague IBA

Nissequogue River Watershed and Smithtown Bay

Brookhaven, Smithtown, Suffolk County

26,000 acres
0-200' elevation

40.9124°N
73.2357°W

IBA Criteria Met

Criterion	*Species*	*Data*	*Season*	*Source*
Species at Risk	Piping Plover	8 pairs in 2001, 4 in 2000, 7 in 1999, 10 in 1998, 10 in 1997, 8 in 1996, 7 in 1995, 3 in 1994, 4 in 1993	Breeding	NY Natural Heritage Biodiversity Databases
Species at Risk	Common Tern	582 pairs in 1999, 495 in 1998, 747 in 1997, 685 in 1996, 766 in 1995, 358 in 1994, 406 in 1993	Breeding	NY Natural Heritage Biodiversity Databases
Species at Risk	Least Tern	107 pairs in 2001, 99 in 2000, 208 in 1999, 228 in 1998, 296 in 1997, 194 in 1996, 139 in 1995, 52 in 1994, 23 in 1993	Breeding	NY Natural Heritage Biodiversity Databases
Congregations-Waterbirds	Terns	Estimated 1 pair in 2002, 107 in 2001, 99 in 2000, 790 in 1999, 723 in 1998, 1,043 in 1997, 879 in 1996, 905 in 1995, 410 in 1994, 429 in 1993	Breeding	NY Natural Heritage Biodiversity Databases
Congregations-Wading Birds	Great Egret, Snowy Egret, and Black-crowned Night-Heron	214 pairs in 2004	Breeding	Mike Wasilco pers. comm. 2004

Description: Located on the north shore of Long Island, this site extends from Sunken Meadow State Park in the west to Crane Neck Point in the east, including the offshore waters of Smithtown Bay, and inland to Blydenburgh County Park. Ownership is a mix of private, municipal, county, and state. Sunken Meadow State Park, Nissequogue River State Park, and Caleb Smith State Park Preserve are administered by NYS OPRHP, while Long Beach and Short Beach are owned by the Town of Smithtown. The site includes sandy/cobble shoreline, small areas of spartina marsh, mudflats, and riparian habitat. Blydenburgh County Park is mainly wooded, with an L-shaped shallow pond.

Birds: This site supports colonial nesting birds, including Piping Plovers, Common Terns, and Least Terns, as well as more than 100 herons and egrets. The site also contains freshwater wintering waterfowl habitat that supports at least 1,500-1,800 individuals (500 Canvasbacks and 500 Ring-necked Ducks were reported in winter 1996 at one location).

Conservation: This site is listed in the 2002 Open Space Conservation Plan as a priority site under the project name Long Island Sound Coastal Area. Management efforts are needed to eliminate or minimize human disturbance and intrusions into nesting colonies of terns and Piping Plovers on beaches during the critical nesting season (mid-April to August) via all means available—including fencing, beach closures, posting, beach warden patrols, and public education. In colonies where predation is a significant problem, whether from pets, feral animals, or native species such as foxes, raccoons, or gulls, predator control programs should be considered. Current ordinances prohibit dogs on beaches during nesting season, and prohibit the use of ATVs. Predator exclosures are constructed around Piping Plover nests and string fencing is installed around Piping Plover and Least Tern nesting areas to minimize human disturbance. Boat landing is prohibited during the nesting season on Young's Island, where most Common Terns nest. These regulations should continue to be enforced. Certain upland portions of the area are subject to increasing pressure for commercial and residential development. A 153-acre portion of the former New York State Kings Park Psychiatric Center has been transferred to NYS OPRHP as the Nissequogue River State Park. The Nissequogue River has been designated as a state Bird Conservation Area. A secluded 3-acre pond at the mouth of the Nissequogue River in the Nissequogue River State Park has served as a roost for wading birds and is currently undergoing habitat restoration. Monitoring of at-risk species and waterfowl use of the area should continue.

North Brother/South Brother Islands

Bronx, Bronx County

28 acres
10' elevation

40.7995°N
73.8993°W

IBA Criteria Met

Criterion	*Species*	*Data*	*Season*	*Source*
Congregations-Wading Birds	Mixed species	Estimated 497 nests in 2004, 619 in 2003, 609 in 2002, 621+ in 2001 [1]; Estimated 168 nesting pairs in 1995, 211 in 1994, 110 in 1993, 95 in 1992 [2]	Breeding	1. NY City Audubon Harbor Herons report; 2. NY Natural Heritage Biodiversity Databases

Description:

This site consists of two currently uninhabited islands in the East River in New York City. Both South Brother and North Brother were inhabited at one time. North Brother has a history of municipal development and several abandoned buildings are present. The islands are a mix of deciduous forest, scrub, and grassy patches. North Brother Island is owned by the City of New York and South Brother Island is privately owned.

Birds:

These islands are important wading bird nesting sites. Surveys in 2004 showed that the islands support breeding Great Egrets (estimated 60 nests), Snowy Egrets (estimated 65 nests), Cattle Egrets (two nest), Black-crowned Night-Herons (365 nests), and Glossy Ibis (four nests). Also present during the 2004 surveys were 350 Double-crested Cormorant, 140-plus Herring Gull, and 90-plus Great Black-backed Gull nests.

Conservation:

This site is listed in the 2002 Open Space Conservation Plan as a priority site under the project name Harbor Herons Wildlife Complex. A partnership project involving Audubon New York, New York City Audubon, and the New York City Department of Parks and Recreation is underway to restore habitat on North Brother Island. South Brother is subject to potential development and should be set aside as a protected natural area. New York City Audubon has surveyed this site as part of their Harbor Herons surveys and monitoring should continue. During the first round of IBA site identifications, the site was recognized under the research criterion because a long-term monitoring project is based there.

Northwest Harbor/ Shelter Island Complex

Multiple municipalities, Suffolk County

24,300 acres
0-100' elevation

41.0407°N
72.3142°W

IBA Criteria Met

Criterion	*Species*	*Data*	*Season*	*Source*
Species at Risk	Osprey	Estimated 25 breeding pairs	Breeding	Mike Scheibel pers. comm. 2003
Species at Risk	Piping Plover	16 pairs in 1999, 24 in 1998, 30 in 1997, 28 in 1996, 28 in 1995, 23 in 1994, 17 in 1993	Breeding	NY Natural Heritage Biodiversity Databases
Responsibility Species Assemblage-Forest	Black-billed Cuckoo, Hairy Woodpecker, Northern Flicker, Eastern Wood-Pewee, Great Crested Flycatcher, Wood Thrush, Black-and-white Warbler, Scarlet Tanager, Baltimore Oriole	Breed	Breeding	NY BBA 2000
Responsibility Species Assemblage-Shrub	Northern Bobwhite, American Woodcock, Eastern Kingbird, Gray Catbird, Brown Thrasher, Blue-winged Warbler, Prairie Warbler, Eastern Towhee, Field Sparrow	Breed	Breeding	NY BBA 2000
Congregations-Waterfowl	Mixed species	The Mashomack and Sag Harbor supported thousands of waterfowl in 1995 [1]; the area is thought to sometimes hold as many as 20,000 to 40,000 waterfowl at peak times [2]	Winter	1. Christmas Bird Count; 2. Nomination, Steven Biasetti
Congregations-Waterbirds	Terns	Estimated 10 pairs in 1999, 31 in 1998, 69 in 1997, 2 in 1996, 165 in 1995, 85 in 1994, 68 in 1993	Breeding	NY Natural Heritage Biodiversity Databases

Piping Plover

Description: This site includes an area of diverse habitats on and around Shelter Island, including Cedar Beach Point on Great Hog Neck, The Nature Conservancy's Mashomack Preserve, Morton National Wildlife Refuge, Grace Estate, and adjoining shoreline and offshore waters. According to the NY GAP land cover data, approximately 25% of this site is pitch pine oak, which supports both forest and shrub species. Another three percent of the site includes deciduous wetland and successional hardwoods. Another five percent of the site includes successional shrub, old/field pasture, and cropland. The waters in this area serve as an important summer feeding and nursery area for juvenile Kemp's ridley sea turtles (*Lepidochelys kempii*), a federally endangered species.

Birds: At-risk species that breed here include Ospreys (estimated 25 pairs), Piping Plovers (16 pairs in 1999), Common Terns (two pairs in 1996, 14 pairs in 1995), and Least Terns (87 pairs in 1996; 3% of state population). Horned Larks (four pairs in 1996) breed in some of the dune areas. Upland mainland portions include Grace Estate, a 514-acre mixed deciduous forest that hosts breeding species unusual to Long Island, such as the Acadian Flycatcher, Veery, Cerulean Warbler (three pairs in 1996), Black-and-white Warbler, American Redstart, Ovenbird, and

Scarlet Tanager. The Nature Conservancy's Mashomack Preserve on Shelter Island is a 2,100-acre, largely unfragmented deciduous forest covering a third of the island. The site is also an important waterfowl wintering area, thought to hold as many as 20,000 to 40,000 waterfowl at peak times. The Mashomack and Sag Harbor portions of the 1995 Orient Christmas Bird Count documented 876 American Black Ducks, 3,000-plus scoters (mostly White-winged), 472 Long-tailed Ducks, 254 Common Goldeneyes, and 205 Red-breasted Mergansers.

Conservation: This site is listed in the 2002 Open Space Conservation Plan as a priority site under the project name Peconic Pinelands Maritime Reserve Project. Within this site, some large preserves in public ownership are managed to benefit birds, including the Mashomack Preserve, Grace Estate, and Morton National Wildlife Refuge. Certain beach areas used by nesting Piping Plovers and Least Terns, such as Cedar Beach County Park, are heavily impacted by the use of recreational and off-road vehicles. Recent efforts to fence and monitor the area have proven successful and should continue. The Nature Conservancy has been supervising a volunteer tern and plover monitoring program in the area. The North Fork Audubon Society has proposed a more active monitoring and management program. This would likely involve not only North Fork Audubon and The Nature Conservancy, but also the Suffolk County Department of Parks and Cornell Cooperative Extension. Residential development is increasing, especially on mainland areas of the South Fork, and may lead to further habitat loss, fragmentation, and increased water pollution. As human populations increase, there is concern that increased recreational use of protected areas may negatively affect birds and other wildlife. Inventory and monitoring of at-risk and other priority birds should continue.

Orient Point and Plum Island

Southold, Suffolk County

10,000 acres
0-100' elevation

41.1423°N
72.2713°W

IBA Criteria Met

Criterion	*Species*	*Data*	*Season*	*Source*
Species at Risk	Piping Plover	1 pair in 2002, 1 in 2001, 7 in 1998, 6 in 1997, 9 in 1996, 5 in 1995, 8 in 1994, 5 in 1993	Breeding	NY Natural Heritage Biodiversity Databases
Species at Risk	Common Tern	At least 40 nesting pairs	Breeding	Mike Wasilco pers. comm. 2004
Species at Risk	Least Tern	27 pairs in 1998, 16 in 1997, 108 in 1996, 23 in 1995, 16 in 1994, 48 in 1993	Breeding	NY Natural Heritage Biodiversity Databases
Congregations-Waterbirds	Terns	300+ Common and Roseate Terns courting and fishing in the area between Plum Island and Orient Point	Breeding and migration	Mike Wasilco pers. comm. 2004

Description: This site includes land and water on the North Fork of Long Island, extending from Orient Harbor to Plum Island and including Orient Beach State Park. Between Orient Point and Plum Island lies Plum Gut, a deep open water channel that links the waters of Gardiners Bay with the waters of eastern Long Island Sound. The habitats of particular significance to birds and other wildlife include barrier beaches, salt marshes, shallow bays, and maritime forests. Plum Island has a mixture of rocky shoreline, sand beaches, wetlands, and various upland shrub, grassland, and forest habitats. Several regionally rare plant species occur here, including Scotch loveage (*Ligusticum scothicum*), slender knotweed (*Polygonum tenue*), and sea-beach knotweed (*Polygonum glaucum*). A stand of blackjack oak (*Quercus marilandica*) represents the northernmost extent of the range of the species. Orient Harbor supports a significant bay scallop (*Aequipecten irradians*) commercial shellfishery and is an important spawning, nursery, and feeding area for a variety of fish. The offshore waters, especially of Plum Gut, host large concentrations of striped bass (*Morone saxatilis*), bluefish (*Pomatomus saltatrix*), tautog (*Tautoga onitis*), summer flounder (*Paralichthys dentatus*), and others. Plum Gut is a major migration corridor for striped bass and Atlantic salmon (*Salmo salar*).

Birds: Colonial breeding birds documented here during the 1995 NYS DEC LICWPP survey included Great Egrets (18 pairs), Snowy Egrets (five pairs), Black-crowned Night-Herons (14 pairs), Piping Plovers (five pairs), American Oystercatchers (five pairs), Herring Gulls (2,608 pairs), Great Black-backed Gulls (1,691 pairs), and Least Terns (23 pairs). There were also 27 pairs of Double-crested Cormorants. Plum Gut, between Orient Point and Plum Island, is a nutrient-rich upwelling that is an important feeding area for Roseate and Common Terns from the nearby Great Gull Island colony. Ospreys nest and forage in the marshes here, and the area is an important waterfowl wintering area with substantial numbers of Canada Geese, American Black Ducks, Mallards, Canvasbacks, scaup, Long-tailed Ducks, scoters, Buffleheads, Common Goldeneyes, and Red-breasted Mergansers.

Conservation: This site is listed in the 2002 Open Space Conservation Plan as a priority site under the project name Long Island Sound Coastal Area. Intensive management efforts are needed to eliminate or minimize human disturbance and intrusions into nesting colonies of terns and Piping Plovers at Orient Point during the critical nesting season (mid-April to August). Means to accomplish this include fencing, beach closures, posting, beach warden patrols, and public education. In those colonies where predation is a significant problem, whether from pets, feral animals, or native species such as raccoons or gulls, predator control programs should be undertaken. NYS OPRHP should continue its stewardship program for Piping Plovers in cooperation with The Nature Conservancy. Management plans should be developed and implemented by state, town, and private conservation groups. Increased development of the shoreline in the Orient Harbor area could degrade water quality and the suitability of these waters and habitats. Monitoring of at-risk species and waterfowl is needed.

Oyster Bay Area

Oyster Bay and Huntington, Nassau and Suffolk Counties

12,700 acres
0-200' elevation

40.9029°N
73.5196°W

IBA Criteria Met

Criterion	*Species*	*Data*	*Season*	*Source*
Species at Risk	American Black Duck	171 ind. in 1999, 260 in 2000	Winter	NYSOA winter waterfowl counts
Congregations-Waterfowl	Mixed species	2,096 ind. in 1996 [1]; many ind. species counts exceed threshold (see text) [2]	Winter	1. NYS DEC mid-winter aerial waterfowl surveys; 2. Oyster Bay National Wildlife Refuge survey

Greater Scaup

Description: This site encompasses the area extending along western Long Island Sound from Bayville eastward to Cold Spring Harbor, including the Oyster Bay National Wildlife Refuge. Much of the shoreline has been developed for residential houses and estates, but many significant wetland, shrub/scrub, and deciduous forest habitats remain. It is mostly under private ownership, but there are significant public holdings within Oyster Bay National Wildlife Refuge, and other municipal lands.

Birds: The NYS DEC aerial winter waterfowl survey in 1996 documented 2,096 waterfowl in this area, including 1,040 Canada Geese. Waterfowl surveys done from 1990 to 1997 at Oyster Bay National Wildlife Refuge showed maximums of 781 Canada Geese, 140 American Wigeons, 2,133 American Black Ducks, 234 Mallards, 1,048 Canvasbacks, 17,102 Greater Scaup, 861 Long-tailed Ducks, 2,082 Buffleheads, 203 Common Goldeneyes, and 322 Red-breasted Mergansers. Non-breeding maximums for other waterbirds during this time included 209 Double-crested Cormorants, 1,262 Herring Gulls, 356 Common Terns, 299 Forster's Terns, and 148 herons and egrets. Upland portions of the site are known for having a large diversity and abundance of migratory songbirds during spring and fall migrations.

Conservation: Water pollution from various sources, including contaminants, oil spills, suburban runoff, excessive sedimentation, and sewage and storm water discharges, degrades the aquatic resources upon which these waterfowl and other waterbirds rely. Also, increased residential and commercial development along the shoreline may cause increased pollution and disturbance problems. Protection of significant areas, particularly wetland sites, should continue. Waterfowl numbers should be regularly monitored.

Peconic Bays and Flanders Bay

Riverhead, Southold, Southampton, Suffolk County

22,800 acres
0-50' elevation

40.9561°N
72.4997°W

IBA Criteria Met

Criterion	*Species*	*Data*	*Season*	*Source*
Species at Risk	Piping Plover	1 pair in 2002, 13 in 1999, 19 in 1998, 23 in 1997, 20 in 1996, 24 in 1995, 16 in 1994, 10 in 1993	Breeding	NY Natural Heritage Biodiversity Databases
Species at Risk	Least Tern	461 pairs in 1999, 291 in 1998, 307 in 1997, 368 in 1996, 254 in 1995, 147 in 1994, 168 in 1993	Breeding	NY Natural Heritage Biodiversity Databases
Congregations-Waterfowl	Mixed species	2,968 ind. in 2000 [1]; 2,051 in 1994 [2]	Winter	1. NYSOA winter waterfowl counts 2. NYS DEC mid-winter aerial waterfowl surveys
Congregations-Waterbirds	Terns	Estimated 2 pairs in 2002, 461 in 1999, 296 in 1998, 311 in 1997, 375 in 1996, 263 in 1995, 156 in 1994, 192 in 1993	Breeding	NY Natural Heritage Biodiversity Databases

Description: This site is a large, sheltered marine bay system located between the North and South Forks of eastern Long Island, extending from the mouth of the Peconic River in the west to Little Peconic Bay in the east, including the adjacent shoreline and Robins Island. Along with the Peconic River, several smaller tributaries, including Goose Creek, Birch Creek, Mill Creek, and Hubbard Creek empty into the bays. The shores of Flanders Bay contain a relatively large (800-acre) undisturbed salt marsh complex. This very productive marine ecosystem serves as an important nursery area for a variety of finfish and shellfish.

Birds: This is an important breeding area for colonial waterbirds, including Piping Plovers (18 pairs in 1996; 7% of the state population), American Oystercatchers (10 pairs in 1996; 6% of the state population), Common Terns (nine pairs in 1996), Least Terns (369 pairs in 1996; 12% of the state population), and Black Skimmers (one pair in 1996). Ospreys nest here and forage in the wetlands. It is also an important wintering

Long-tailed Ducks

and staging area for waterfowl, loons, and grebes, particularly Canada Geese, American Black Ducks, scaup, Long-tailed Ducks, Red-breasted Mergansers, and Common Loons. The NYS DEC aerial waterfowl survey tallied 2,051 waterfowl in January 1994.

Conservation: This site is listed in the 2002 Open Space Conservation Plan as a priority site under the project name Peconic Pinelands Maritime Reserve Project. Inputs of nutrients into Peconic Bay from both point and non-point sources need to be greatly reduced. Degradation of water quality, especially by non-point source runoff, is a concern. Non-point sources released into the bay produce nutrient loading, followed by eutrophication and increased levels of fecal bacteria. Public outreach efforts should focus on reducing fertilizer and pesticide use, and proper maintenance of septic systems by landowners. Disturbances to colonial beach-nesting birds and wintering waterfowl should be minimized. Nesting terns and Piping Plovers, in particular, could benefit from programs involving public education, warden patrols, protective exclosures, predator control, and beach closures. Local conservation groups should consider negotiation of conservation easements or management agreements for some important privately owned beaches. Monitoring of breeding colonial waterbirds should continue, and a better system for monitoring wintering waterfowl and public use of the area should be implemented.

Pelham Bay Park

Bronx, Bronx County

2,700 acres
0-100' elevation

40.8724°N
73.8040°W

IBA Criteria Met

Criterion	*Species*	*Data*	*Season*	*Source*
Species at Risk	Least Bittern	Site is in need of further monitoring to document numbers of at-risk species	Breeding	IBA nomination, David Künstler

Description: Pelham Bay Park is located on the shore of Long Island Sound and is the largest natural area complex within the region. The park has a wide range of habitats, including forests (782 acres), meadows (83 acres), mixed scrub (51 acres), salt marsh (195 acres), fresh water marsh (three acres), salt flats (161 acres), and a stretch of saltwater coastline. The site also includes a golf course, parking lot, human-made beach, and re-vegetating landfill. The site supports a number of state-listed plant species, including wild pink (*Silene caroliniana*), slender blue flag (*Iris prismatica*), field beadgrass (*Paspalum laeve*), slender spikerush (*Eleocharis tenuis var. pseudoptera*), persimmon (*Diospyros virginiana*), and gamma grass (*Tripsacum dactyloides*). The site is the only known location in the state that supports a species of noctuid moth (*Amphipoea erepta ryensis*), which is dependent on gamma grass. This site is owned by the New York City Department of Parks and Recreation.

Birds: As one of few large natural areas in the heavily urbanized New York City region, the site is particularly important to the remaining breeding forest and wetland bird species. It also serves as a stopover site for a great diversity of migrant birds. The deciduous forest and scrub habitats support a characteristic breeding community, including the Red-eyed Vireo, Wood Thrush, Gray Catbird, Chestnut-sided Warbler, Ovenbird, Common Yellowthroat, and Eastern Towhee. The salt marshes support the breeding Clapper Rail (1 pair in 1994), Marsh Wren (20 pairs in 1994), Saltmarsh Sharp-tailed Sparrow (six pairs in 1994), and Swamp Sparrow (15 pairs in 1994). The site has a small rookery with Snowy Egrets (two pairs in 1996), Black-crowned Night-Herons (27 pairs in 1996), and Yellow-crowned Night-Herons (one pair in 1996). The salt marshes are also used extensively for foraging by herons and egrets from nearby rookeries. Occasional breeders at the site include the Least Bittern (one pair in 1989), Least Tern, and Barn Owl. The site is also a well-known fall hawk watch location, with season totals of 21,951 hawks in 1990 and 12,065 hawks in 1988. The hawk watch at Pelham Bay regularly has tallied one of the highest Osprey counts in the U.S. (third highest in 1988 and 1989).

Conservation: Competition from non-native plants is negatively affecting many plant communities in the park, and may be impacting bird abundance. The presence of small culverts under some roads restricts the flow of the tidal marshes and may decrease overall ecosystem health, which may partly account for decreased densities of salt marsh bird species (121 pairs of Saltmarsh Sharp-tailed Sparrows in 1958 as compared to six pairs in 1994). The New York City Department of Parks and Recreation completed a natural areas management plan for the park in 1988, which recommended removal of non-native plants, trail restoration, pond creation, and amphibian and reptile reintroductions. More inventories and monitoring of at-risk and migrating species are needed to support the IBA status of this site.

Pelham Bay Park

Prospect Park

Brooklyn, Kings County

526 acres
70-150' elevation

40.6614°N
73.9696°W

IBA Criteria Met

Criterion	*Species*	*Data*	*Season*	*Source*
Congregations-Migrant Landbirds	Mixed species	Site supports an exceptional diversity of migrant songbirds and may be an important migratory stopover for landbirds. It is common in spring to see 100 or more species in one day, often including at least 25 species of warblers.	Migration	Nomination, Paul Keim

Description: This urban park is a designed landscape built by the famous park designers Olmsted and Vaux in the late 1860s. It combines elements of a romantic English pastoral landscape (trees and turf), with surviving northern deciduous forest on outwash and glacial moraine soils. The park also contains human-made small ponds (each less than an acre), a lake (55 acres), and managed wetlands. Approximately half of the park is wooded. A forested island in the middle of a highly populated city, the park harbors the last remaining eastern deciduous forest in Brooklyn, and as such, offers New York City residents an educational opportunity.

Birds: The site supports an exceptional diversity of migrant songbirds and is thought to be an important migratory stopover for landbirds. More than 240 species have been recorded in the park, and it is not uncommon in spring to see 100 or more species in one day. Often, 25 or more of these species are warblers. The site also regularly harbors five to 15 wintering Pied-billed Grebes among a large diversity of waterfowl. The park's relatively large size and position as an "island" of forest and wetland habitat amidst a sea of nearly complete urbanization makes it particularly important as one of the few regional sites supporting characteristic breeding and migrant bird species.

Magnolia Warbler

Conservation: Over seven million annual visitors to Prospect Park make human disturbance a great threat. Soil compaction and erosion and increases in detrimental invasive plant species stem from human disturbance. Large numbers of introduced bird species (especially European Starlings and House Sparrows) compete with native species. A 25-year management plan has been developed for the natural areas of the park. Included in the plan are proposals to combat soil erosion, replant habitats with native species, create and manage wildlife habitats, and increase the diversity and abundance of invertebrate species. Funding to implement the plan has been secured through not-for-profit fundraising efforts. Much of the initial restoration recommended in the plan to increase diversity within the ecosystem has been carried out, including plantings that provide nutrition and shelter for the wide variety of birds supported at the site.

Shinnecock Bay

Shinnecock and Southampton, Suffolk County

12,500 acres
0-50' elevation

40.8406°N
72.4971°W

IBA Criteria Met

Criterion	*Species*	*Data*	*Season*	*Source*
Species at Risk	Piping Plover	3 pairs in 2001 13 in 1999, 17 in 1998, 14 in 1997, 19 in 1996, 13 in 1995, 15 in 1994, 8 in 1993	Breeding	NY Natural Heritage Biodiversity Databases
Species at Risk	Roseate Tern	7 pairs in 1999, 34 in 1998, 59 in 1997, 95 in 1996, 95 in 1995, 55 in 1994, 45 in 1993	Breeding	NY Natural Heritage Biodiversity Databases
Species at Risk	Common Tern	484 pairs in 1999, 1,414 in 1998, 1,406 in 1997, 1,723 in 1996, 1,417 in 1995, 1,413 in 1994, 1,898 in 1993	Breeding	NY Natural Heritage Biodiversity Databases
Species at Risk	Least Tern	183 pairs in 1999, 214 in 1998, 278 in 1997, 187 in 1996, 230 in 1995, 65 in 1994, 76 in 1993	Breeding	NY Natural Heritage Biodiversity Databases
Species at Risk	Black Skimmer	0 pairs in 1999, 19 in 1998, 24 in 1997, 15 in 1996, 28 in 1995, 58 in 1994, 55 in 1993	Breeding	NY Natural Heritage Biodiversity Databases
Responsibility Species Assemblage-Wetland	American Black Duck, Glossy Ibis, Clapper Rail, Marsh Wren, Saltmarsh Sharp-tailed Sparrow, Seaside Sparrow	Breed	Breeding	NY BBA 2000
Congregations-Waterfowl	Mixed species	The site is one of the most important waterfowl wintering areas on Long Island, with average concentrations of 3,500 birds and 7,284 in the peak year	Winter	NYS DEC mid-winter aerial waterfowl surveys

Criterion	*Species*	*Data*	*Season*	*Source*
Congregations-Wading Bird	Mixed species	Estimated 365 pairs in 1995, 179 in 1998	Breeding	Long Island Colonial Waterbird and Piping Plover Survey
Congregations-Waterbirds	Terns	Estimated 674 pairs in 1999, 1,662 in 1998, 1,743 in 1997, 2,005 in 1996, 1,742 in 1995, 1,533 in 1994, 2,019 in 1993	Breeding	NY Natural Heritage Biodiversity Databases
Congregations-Individual Species	Great Egret	110 pairs in 1995; 21% of state population	Breeding	Long Island Colonial Waterbird and Piping Plover Survey
Congregations-Individual Species	Snowy Egret	45 pairs in 1995; 7% of state population	Breeding	Long Island Colonial Waterbird and Piping Plover Survey
Congregations-Individual Species	Black-crowned Night-Heron	150 pairs in 1995; 8% of state coastal population	Breeding	Long Island Colonial Waterbird and Piping Plover Survey
Congregations-Individual Species	Glossy Ibis	60 pairs in 1995; 8% of state population	Breeding	Long Island Colonial Waterbird and Piping Plover Survey
Congregations-Individual Species	Piping Plover	20 pairs in 1996; 8% of state population	Breeding	Long Island Colonial Waterbird and Piping Plover Survey
Congregations-Individual Species	American Oystercatcher	36 pairs in 1996; 20% of state population	Breeding	Long Island Colonial Waterbird and Piping Plover Survey
Congregations-Individual Species	Herring Gull	800 pairs in 1995; 7% of state population	Breeding	Long Island Colonial Waterbird and Piping Plover Survey
Congregations-Individual Species	Great Black-backed Gull	550 pairs in 1995; 8% of state population	Breeding	Long Island Colonial Waterbird and Piping Plover Survey
Congregations-Individual Species	Roseate Tern	95 pairs in 1996; 6% of state population	Breeding	Long Island Colonial Waterbird and Piping Plover Survey
Congregations-Individual Species	Common Tern	1723 pairs in 1996; 9% of state coastal population	Breeding	Long Island Colonial Waterbird and Piping Plover Survey
Congregations-Individual Species	Least Tern	179 pairs in 1996; 6% of state population	Breeding	Long Island Colonial Waterbird and Piping Plover Survey
Congregations-Individual Species	Black Skimmer	25 pairs in 1996; 5% of state population	Breeding	Long Island Colonial Waterbird and Piping Plover Survey

Description: This site includes a diverse region of barrier island beaches, salt marshes, dredge spoil islands, and surrounding bays and estuaries. It includes five miles of mostly undeveloped shoreline along the southernmost part of Shinnecock Bay, and large undeveloped tidal wetlands that are relatively rare in the state. The site is owned by municipalities and private landowners. According to the NY GAP land cover data, the site includes 710 acres of salt marsh habitat. Diamondback terrapins (*Malaclemys terrapin*) use inaccessible salt marsh islands for nesting.

Black Skimmer

Birds: The area supports significant numbers of colonial nesting birds, including Great Egrets (110 pairs in 1995; 21% of state population), Snowy Egrets (45 pairs in 1995; 7% of state population), Black-crowned Night-Herons (150 pairs in 1995; 8% of state coastal population), Glossy Ibis (60 pairs in 1995; 8% of state population), Piping Plovers (20 pairs in 1996; 8% of state population), American Oystercatchers (36 pairs in 1996; 20% of state population), Herring Gulls (800 pairs in 1995; 7% of state population), Great Black-backed Gulls (550 pairs in 1995; 8% of state population), Roseate Terns (95 pairs in 1996; 6% of state population), Common Terns (1,723 pairs in 1996; 9% of state coastal population), Least Terns (179 pairs in 1996; 6% of state population), and Black Skimmers (25 pairs in 1996; 5% of state population). The site is one of the most important waterfowl wintering areas on Long Island, with average concentrations of 3,500 birds from 1975 to 1984, and 7,284 in the peak year. This included an average of 470 Brant (maximum 1,060), 1,650 scaup (maximum 4,100), 380 American Black Duck (maximum 867), 300 Bufflehead (maximum 1,265), 100 Common Goldeneye (maximum 305), and 400 Red-breasted Merganser (maximum 1,455). Salt marsh habitats support a characteristic breeding bird community, including the American Black Duck, Glossy Ibis, Clapper Rail, Marsh Wren, Saltmarsh Sharp-tailed Sparrow, and Seaside Sparrow. It is also an important coastal hawk migration corridor and supports wintering Northern Harriers and Short-eared Owls.

Conservation: This site is listed in the 2002 Open Space Conservation Plan as a priority site under the project name Long Island South Shore Estuary Preserve. Portions of this site have been designated as a state Bird Conservation Area. Several ecologically critical areas along the bay side of the barrier island are at risk due to potential development. Some residential and commercial developments have already occurred on ecologically important lands. Other potential threats include water quality degradation, unregulated dredge spoil disposal, dock construction, and increased recreational uses. Levels of human disturbance to colonial nesting birds should be monitored to prevent losses of eggs or young. Inventory and monitoring, particularly of at-risk species, should continue.

Southampton Green Belt

Southampton, Suffolk County

5,900 acres
0-245' elevation

40.9443°N
72.3713°W

IBA Criteria Met

Criterion	*Species*	*Data*	*Season*	*Source*
Responsibility Species Assemblage-Forest	Black-billed Cuckoo, Hairy Woodpecker, Northern Flicker, Eastern Wood-Pewee, Great Crested Flycatcher, Wood Thrush, Scarlet Tanager, Rose-breasted Grosbeak, Baltimore Oriole	Breed	Breeding	NY BBA 2000

Description: This site includes approximately 10 to 12 relatively large forest patches, extending from Tuckahoe (and including Tuckahoe Woods) in the west to the Bridgehampton/Sag Harbor area in the east. It is a biologically diverse area of low, rolling oak-heath woods and mixed forests, punctuated by glacial kettles, vernal ponds, and red maple-hardwood swamps. According to the NY GAP land cover data, approximately 90% of this site is forest habitat, which includes oak, pitch-pine oak, and successional hardwoods. Wetlands support the eastern spadefoot toad (*Scaphiopus holbrookii*), which is uncommon on Long Island.

Birds: This site consists of some of the few remaining large forest patches in eastern Long Island. The Acadian Flycatcher and Cerulean Warbler are documented breeders here, occurring at few other locations on Long Island. Other species characteristic of deciduous and mixed woods that breed here include the Black-billed Cuckoo, Hairy Woodpecker, Northern Flicker, Eastern Wood-Pewee, Great Crested Flycatcher, Wood Thrush, Scarlet Tanager, Rose-breasted Grosbeak, and Baltimore Oriole.

Baltimore Oriole

Conservation: This site is listed in the 2002 Open Space Conservation Plan as a priority site under the project name Peconic Pinelands Maritime Reserve Project. Despite their importance, these forest patches are vulnerable and largely unprotected. Several large residential development projects, which would cut deeply into the area's forests, have been proposed. A coalition of grassroots environmental and civic groups and hundreds of area residents have been working hard to increase protection for Tuckahoe Woods. The Town of Southampton owns about 25 acres of the site and has targeted another 45 acres as an immediate acquisition priority for the town. The Peconic Land Trust and The Group for the South Fork are working actively to further land protection efforts in the area. Inventory and monitoring of forest interior species—particularly at-risk species such as the Cerulean Warbler—are needed.

Van Cortlandt Park

Bronx and Yonkers,
Bronx and Westchester Counties

1,135 acres
40-200' elevation

40.8969°N
73.8866°W

IBA Criteria Met

Criterion	*Species*	*Data*	*Season*	*Source*
Congregations-Migrant Landbirds	Mixed species	Supports an exceptional diversity of migrant songbirds and is thought to be an important migratory stopover for landbirds. Nearly every species of landbird in the region has been recorded in the Croton Woods portion of the park without extraordinary effort.	Migration	Nomination, David Künstler

Description: This site is the third largest park in New York City and contains some of the largest areas of natural land in the city. The park is surrounded on three sides by urbanization with several natural corridors extending north into Westchester County on one side. Most of the northern half of the park is quality deciduous forest, including the 158-acre Croton Woods. The park also contains several streams, ponds, and marshes. In the southern portion of the park there are two golf courses and numerous sports fields and recreational areas. The site hosts a number of unusual plant species, including wild pink (*Silene caroliniana),* purple milkweed (*Asclepias purpurascens*), goldenseal (*Hydrastis canadensis*), Schrener's aster (*Eurybia schreberi*), wingstem (*Verbesina alternifolia*), field beadgrass (*Paspalum laeve*), and Bicknell's sedge (*Carex bicknelli*). The park also supports some of the largest known populations of two North American butterflies, the silvery checkerspot (*Chlosyne nycteis*) and the hoary edge (*Achalarus lyciades*).

Birds: Van Cortlandt Park supports an exceptional diversity of migrant songbirds and is thought to be an important migratory stopover for landbirds. In 1987, 153 species were recorded in the park, most of which were landbirds. Nearly every species of landbird in the region has been recorded in the Croton Woods portion of the park, without any extraordinary effort. The park's relatively large size and position as an "island" of quality forest and wetland habitat in a sea of nearly complete urbanization makes it particularly important as one of the last regional sites supporting characteristic breeding and migrant bird species. These include the Broad-winged Hawk, Hairy Woodpecker, Northern Flicker, Eastern Wood-Pewee, Great Crested Flycatcher, Wood Thrush, Rose-breasted Grosbeak, and Baltimore Oriole.

Conservation: Competition from non-native plants is an ongoing problem, but a program to eliminate or decrease the abundance of non-natives was completed in 1996. Succession threatens a high-quality little bluestem (*Schizachyrium scoparium*) grassland if burning is prevented. The woodland areas that provide some of the most important habitat for migrant songbirds are negatively impacted by vandalism and recreational overuse. Future management plans should recognize the importance of this island of woodland habitat to birds that rely on the park for breeding and migration.

West Hempstead Bay/ Jones Beach West

Multiple municipalities, Nassau and Suffolk County

32,000 acres
0-10' elevation

40.5975°N
73.5913°W

IBA Criteria Met

Criterion	*Species*	*Data*	*Season*	*Source*
Species at Risk	Brant	19,536 ind in 1999, 6,268 in 2000, 2,980 in 2001	Breeding	NYSOA winter waterfowl counts
Species at Risk	American Black Duck	4,706 ind in 1999, 3,459 in 2000, 3,460 in 2001	Breeding	NYSOA winter waterfowl counts
Species at Risk	Piping Plover	1 pair in 2002, 15 in 2001, 17 in 2000, 38 in 1999, 45 in 1998, 52 in 1997, 49 in 1996, 49 in 1995, 43 in 1994, 48 in 1993	Breeding	NY Natural Heritage Biodiversity Databases
Species at Risk	American Oystercatcher	2 confirmed atlas blocks [1]; hundreds breeding and in migration [2]	Breeding	1. NY BBA 1980 and 2000; 2. Sy Schiff pers. comm. 2004
Species at Risk	Common Tern	93 pairs in 2002, 74 in 2001, 72 in 2000, 1,110 in 1999, 1,468 in 1998, 1,377 in 1997, 1,411 in 1996, 1,387 in 1995, 1,320 in 1994, 1,554 in 1993	Breeding	NY Natural Heritage Biodiversity Databases
Species at Risk	Least Tern	24 pairs in 2002, 10 in 2001, 13 in 2000, 418 in 1999, 202 in 1998, 457 in 1997, 879 in 1996, 587 in 1995, 335 in 1994, 227 in 1993	Breeding	NY Natural Heritage Biodiversity Databases
Species at Risk	Black Skimmer	79 pairs in 1999, 87 in 1998, 94 in 1997, 32 in 1996, 38 in 1995, 24 in 1994, 10 in 1993	Breeding	NY Natural Heritage Biodiversity Databases
Responsibility Species Assemblage-Beach/Dune	Piping Plover, American Oystercatcher, Common Tern, Least Tern, Black Skimmer	Breed	Breeding	NY BBA 2000

Criterion	*Species*	*Data*	*Season*	*Source*
Congregations-Waterfowl	Mixed species	893 ind. in 2004, 8,664 in 2003, 14,644 in 2002, 8,180 in 2001, 17,727 in 2000, 30,961 in 1999 [1]; Possibly more than 5,000 individuals use the site [2]	Winter	1. NYSOA winter waterfowl counts 2. Nomination, Catherine Brittingham
Congregations-Waterbirds	Terns	117 pairs in 2002, 84 in 2001, 85 in 2000, 1,528 in 1999, 1,682 in 1998, 1,840 in 1997, 2,299 in 1996, 1,976 in 1995, 1,658 in 1994, 1,784 in 1993	Breeding	NY Natural Heritage Biodiversity Databases
Congregations-Wading Birds	Mixed species	Estimated 491 nesting pairs in 1995, not surveyed 1994, 422 in 1993, 531 in 1992	Breeding	NY Natural Heritage Biodiversity Databases
Congregations-Individual Species	Great Egret	145 pairs in 1995; 28% of state coastal population	Breeding	Long Island Colonial Waterbird and Piping Plover Survey
Congregations-Individual Species	Snowy Egret	204 pairs in 1995, 30% of state population	Breeding	Long Island Colonial Waterbird and Piping Plover Survey
Congregations-Individual Species	Little Blue Heron	18 pairs in 1995; 52% of state population	Breeding	Long Island Colonial Waterbird and Piping Plover Survey
Congregations-Individual Species	Tri-colored Heron	14 pairs in 1995; 54% of state population	Breeding	Long Island Colonial Waterbird and Piping Plover Survey
Congregations-Individual Species	Black-crowned Night-Heron	209 pairs in 1995; 12% of state coastal population	Breeding	Long Island Colonial Waterbird and Piping Plover Survey
Congregations-Individual Species	Yellow-crowned Night-Heron	13 pairs in 1995; 48% of state population	Breeding	Long Island Colonial Waterbird and Piping Plover Survey
Congregations-Individual Species	Glossy Ibis	275 pairs in 1995; 36% of state population	Breeding	Long Island Colonial Waterbird and Piping Plover Survey
Congregations-Individual Species	Piping Plover	43 pairs in 1996; 1% or more of east coast population, 17% of state population	Breeding	Long Island Colonial Waterbird and Piping Plover Survey
Congregations-Individual Species	American Oystercatcher	46 pairs in 1995; 26% of state population	Breeding	Long Island Colonial Waterbird and Piping Plover Survey

Criterion	Species	Data	Season	Source
Congregations-Individual Species	Herring Gull	2,094 pairs in 1995; 19% of state population	Breeding	Long Island Colonial Waterbird and Piping Plover Survey
Congregations-Individual Species	Great Black-backed Gull	143 pairs in 1995; 2% of state population	Breeding	Long Island Colonial Waterbird and Piping Plover Survey
Congregations-Individual Species	Common Tern	1,404 pairs in 1996; 7% of state coastal population	Breeding	Long Island Colonial Waterbird and Piping Plover Survey
Congregations-Individual Species	Least Tern	927 pairs in 1996; 30% of state population	Breeding	Long Island Colonial Waterbird and Piping Plover Survey
Congregations-Individual Species	Black Skimmer	32 pairs in 1996; 7% of state population	Breeding	Long Island Colonial Waterbird and Piping Plover Survey

Description: This site consists of the barrier islands on the south shore of Long Island, and islands and marshes on the bay side. Sandy beach and dune systems, natural salt marshes and spoil islands are included. According to the NY GAP land cover data, approximately four percent of this site is beach/dune habitat. The site includes Lawrence Marsh, Jones Beach State Park, Point Lookout, Line Islands, and Tobay Beach Park. The interior of the barrier island is bisected by a four-lane highway and heavily developed recreational areas with large parking areas. This site is one of only 12 sites in the state with a population of federally endangered seabeach amaranth (*Polygonum glaucum*).

Birds: Large numbers of waterfowl use this area in winter. The Christmas Bird Count in 1990 documented 25,000 Brant and 10,000 American Black Ducks. Colonial nesting birds include Great Egrets (145 pairs in 1995; 28% of state coastal population), Snowy Egrets (204 pairs in 1995; 30% of state population), Little Blue Herons (18 pairs in 1995; 52% of state population), Tricolored Herons (14 pairs in 1995; 54% of state population), Green Herons (16 pairs in 1995), Black-crowned Night-Herons (209 pairs in 1995; 12% of state coastal population), Yellow-crowned Night-Herons (13 pairs in 1995; 48% of state population), Glossy Ibis (275 pairs in 1995, 36% of state population), Piping Plovers (43 pairs in 1996, 1% or more of east coast population, 17% of state population), American Oystercatchers (46 pairs in 1995, 26% of state population, hundreds are found breeding and during migration), Herring Gulls (2,094 pairs in 1995, 19% of state population), Great Black-backed Gulls (143 pairs in 1995, 2% of state population), Gull-billed Terns (3 in 1996), Common Terns (1,404 pairs in 1996, 7% of state coastal population, not as many in recent years), Forster's Terns (4 pairs in 1996), Least Terns (927 pairs in 1996, 30% of state population), and Black Skimmers (32 pairs in 1996, 7% of state population). The saltmarsh habitats support at-risk species including Brant (confirmed breeder), American Black Duck (confirmed breeder), Common Loon (winters), Pied-billed Grebe (winters), American Bittern (breeds), Osprey (breeds and migrant), Northern Harrier (breeds, winters, and migrant), Sharp-shinned Hawk (migrant), Peregrine Falcon (migrant), Piping Plover (confirmed breeder), American Oystercatcher (confirmed breeder), Common Tern (confirmed breeder), Least Tern (confirmed breeder), Black Skimmer (confirmed breeder), Short-eared Owl (winters), Willow Flycatcher (probable breeder), Saltmarsh Sharp-tailed Sparrow (breeds), and Seaside Sparrow (breeds). The area is important as a migratory shorebird feeding area. Along with other south shore barrier beach islands, this area is a major spring and fall migration route for passerines, accipiters, and falcons.

Conservation: This site is listed in the 2002 Open Space Conservation Plan as a priority site under the project name Long Island South Shore Estuary Preserve. Portions of this site have been designated as a state Bird Conservation Area. Jones Beach is one of the world's most heavily used recreational areas (10 million visitors per year) yet it still maintains significant natural habitats. Intensive management efforts are needed to eliminate or minimize human disturbance and intrusions into nesting colonies of terns and Piping Plovers at beaches during the critical nesting season (mid-April to August), using all available means to accomplish this including fencing, beach closures, posting, beach warden patrols and public education. In those colonies where predation is a significant problem, whether from pets, feral animals or certain problem native species such as raccoons or gulls, removal programs should be undertaken. NYS OPRHP oversees and coordinates a substantial plover stewardship program in cooperation with resource agencies and interest groups. The program includes monitoring, installation of exclosures, and placement of string fencing. There is a need to provide continuing support for this stewardship program. Proposals to improve recreational uses and amenities at the park require close evaluation with respect to potential impacts on birds and bird habitat. The NYS DEC coordinates an annual survey of Piping Plovers and colonial waterbirds. The extensively planted black pines along the barrier beach islands are dying and should be replaced with native salt tolerant species that provide food and shelter to migrant songbirds.

Sources

Able, K.P. 1991. Region 10-Marine Report. Kingbird 41 (4): 281.

Adirondack Cooperative Loon Program. 2004. Annual Census. http://www.adkscience.org/loons/index.htm

Askildsen, J.P. 1993. Region 9-Hudson-Delaware Report. Kingbird 43 (2): 157.

Audubon NY fall surveys at Montezuma: Audubon New York. 2001-2004. Fall shorebird surveys at Montezuma Wildlife Refuge. Unpublished data. Ithaca, New York.

Audubon NY statewide IBA field surveys: Audubon New York. 2003. State-wide Important Bird Areas field surveys. Unpublished data. Ithaca, New York.

BOS noteworthy records: Buffalo Ornithological Society. 2004. Noteworthy Records. http://www.bosbirding.org/counts.htm#noteworthyrecords

Braddock Bay Bird Observatory. 2004. Banding Summary. http://www.bbbo.org

Braddock Bay Raptor Research. 2004. Hawkwatch. http://www.bbrr.org

Brown, K.M., J.L. Tims, R.M. Erwin, and M.E. Richmond. 2001. Changes in the nesting populations of colonial waterbirds in Jamaica Bay Wildlife Refuge, NY, 1974-1998. Northeastern Naturalist 8 (3): 275–292.

BSC Marshbird Monitoring Program: Bird Studies Canada. 2002. The Marshbird Monitoring Program data, 1995-2001. Bird Studies Canada National Data Center, Port Rowan, Ontario.

Burger, M.F. 2000. Breeding Birds of the Montezuma Wetlands Complex: Results of a Pilot IBA Monitoring Effort. Report to the New York Natural Heritage Program. Audubon New York, Ithaca, New York.

Cerulean Warbler Atlas Project: Rosenberg, K.V., S.E. Barker, and R.W. Rohrbaugh. 2000. An Atlas of Cerulean Warbler Populations: Final Report to USFWS: 1997-2000 Breeding Seasons. Cornell Lab of Ornithology, Ithaca, New York.

Christmas Bird Count: National Audubon Society. 2002. The Christmas Bird Count Historical Results [Online]. http://www.audubon.org/bird/cbc

Derby Hill Bird Observatory. 2004. http://www.derbyhill.org

Eagle Institute and Sullivan County Audubon Society. 2002. Winter field surveys of Bald Eagles on the Delaware River. Unpublished data. Barryville, New York.

Harbor Watch at Dunkirk Harbor: Lake Erie Bird Club. Harbor Watch at Dunkirk Harbor. October-March 1982-1989. Daily census of Dunkirk Harbor conducted by members of the Lake Erie Bird Club. Fredonia, New York.

Hawk Migration Association of North America. 2004. http://www.hmana.org/

International Shorebird Surveys (ISS): Manomet Observatory for Conservation Sciences, Manomet, Massachusetts. Contact: Brian Harrington.

Johnson, G. and R.E. Chambers. 1990. Response to roadside playback recordings: an index of red-shouldered hawk breeding density. Pages 71-76 in Ecosystem management: rare species and significant habitats, Proc.15th Ann. Natural Areas Conference.

Johnson, G. and R.E. Chambers. 1994. Productivity and nest success of red-shouldered hawks in central New York. Kingbird 44 (2): 87-95.

Lambert, J.D. 2003. Mountain Birdwatch 2002 Final Report to the United States Fish and Wildlife Service. Vermont Institute of Natural Science, Woodstock, Vermont.

Long Island Colonial Waterbird and Piping Plover Survey: New York State Department of Environmental Conservation, Division of Fish and Wildlife. Long Island Colonial Waterbird and Piping Plover Survey Research Reports, 1992-1999. Stony Brook, New York.

Mazzocchi, I.M. and M. Roggie. 2004. Black Tern (*Chlidonias niger*) Statewide Survey and Observations 2001. New York State Department of Environmental Conservation, Division of Fish, Wildlife and Marine Resources, Watertown, New York.

McCullough R.D., J.F. Farquhar, and I.M. Mazzocchi. 2004. Cormorant Management Activities in Lake Ontario's Eastern Basin. New York State Department of Environmental, Conservation Division of Fish, Wildlife and Marine Resources, Watertown, New York.

NY BBA 1980: New York State Department of Environmental Conservation. New York State Breeding Bird Atlas, 1980-85. Numbers refer to the number of blocks in which a species was found. Definitions of abbreviations in data column of Criteria Met tables: "CO"= confirmed breeding, "PR"= probable breeding, and "PO" = possible breeding. http://www.dec.state.ny.us/apps/bba/results/

NY BBA 2000: New York State Department of Environmental Conservation. New York State Breeding Bird Atlas, 2000-2003. Numbers refer to the number of blocks in which a species was found. The 2000-2003 data were under review during the time of writing; any records are subject to revision or removal. Definitions of abbreviations in data column of Criteria Met tables: "CO"= confirmed breeding, "PR"= probable breeding, and "PO" = possible breeding. http://www.dec.state.ny.us/apps/bba/results/

NY City Audubon Harbor Herons report: New York City Audubon. 2004. Harbor Herons Nesting Survey: 2004. Prepared by Paul Kerlinger, Ph.D. New York City Audubon, New York City, New York.

New York Cooperative Fish and Wildlife Research Unit and Cornell Institute for Resource Information Systems. 1998. The New York Gap Analysis Project Home Page, Version 98.05.01. Department of Natural Resources and Department of Soil, Crop, and Atmospheric Sciences, Cornell University, Ithaca, New York.

New York Cooperative Fish and Wildlife Research Unit. Bird Surveys at Oneida Lake. Departments of Natural Resources and Cornell Biological Field Station. Cornell University, Ithaca, New York. Contact: Milo Richmond, Unit Leader.

NY Natural Heritage Biodiversity Databases: New York Natural Heritage Program, New York State Department of Environmental Conservation. July 2003. Biodiversity Databases, Element Occurrence Record Digital Data Set. Albany, New York.

New York State Department of Environmental Conservation. Hudson River Waterfowl Surveys 2002-2003, Maynard Vance. Albany, New York.

NYS DEC Bald Eagle Report: New York State Department of Environmental Conservation. New York State Bald Eagle Report 2004. Albany, New York.

NYS DEC mid-winter aerial waterfowl surveys: New York State Department of Environmental Conservation. Mid-winter aerial waterfowl surveys, 1973-1994. Unpublished data. Albany, New York.

NYS DEC surveys, Mark Kandel: New York State Department of Environmental Conservation, Region 9. Unpublished bird surveys. Contact: Mark Kandel, Allegany.

New York State Department of State. Undated. Significant Coastal Fish and Wildlife Habitat narratives. Albany, New York. http://nyswaterfronts.com/index.asp

New York State Office of Parks, Recreation and Historic Preservation. 1995. Birds of Fahnestock State Park; A Checklist. Environmental Management Bureau, Albany, New York.

NYSOA winter waterfowl counts: New York State Ornithological Association. 2004. Winter Waterfowl Counts. http://www.nybirds.org/ProjWaterfowl.htm

Oyster Bay National Wildlife Refuge surveys. 1990-1997. Peak Number of Select Waterfowl and Waterbird Species at the Oyster Bay national Wildlife Refuge, a unit of the Long Island National Wildlife Refuge Complex. Unpublished data. Shirley, New York.

Pelkowski, P.I. 2001. Final Report and Summary of Canon 2001 Avian Surveys. Theodore Roosevelt Sanctuary and Audubon Center, Oyster Bay, New York.

Richards, Z. and D. Capen. 2001. An Inventory of Great Blue Heron Rookery on Valcour Island, 2001. University of Vermont, Burlington, Vermont.

Richards, Z. and D. Capen. 2004. Great Blue Herons in the Lake Champlain Ecosystem: An Assessment of the Rookeries on Shad Island, VT and Valcour Island, NY. Draft document. University of Vermont, Burlington, Vermont.

Rimmer, C.C. and K.P. McFarland. 1997. Population Density, Demographics and Distribution of Bicknell's Thrush and other Subalpine Birds on Hunter and Plateau Mountains, New York. Project Report. Vermont Institute of Natural Science, Woodstock, Vermont.

Smith, C.R., S.D. DeGloria, M.E. Richmond, S.K. Gregory, M. Laba, S.D. Smith, J.L. Braden, E.H. Fegraus, J.J. Fiore, E.A. Hill, D.E. Ogurcak, and J.T. Weber. 2001. The New York GAP Analysis Project Final Report. New York Cooperative Fish and Wildlife Research Unit and Cornell Department of Natural Resources, Cornell University, Ithaca, New York.

Smith, G.A. 1989. Point Peninsula, Jefferson County; Mouse raptor mecca during the winter of 1987-1988. Kingbird 39:7-20.

Stevens, G. 1992. Wetlands on the Galeville Army Training Site: report to the United States Military Academy (West Point). Hudsonia Ltd., Bard College Field Station, Annandale, New York.

Sullivan County Audubon Society. 2004. Database of county bird records, 1972-2004. Unpublished data. Loch Sheldrake, New York.

Technical Committee consensus: Upon discussion and debate among the IBA Technical Committee, the extent of habitat and Breeding Bird Atlas presence or other data strongly suggested that the threshold is being met.

The Ripley Hawk Watch. 1993-2003. Unpublished data. Ripley, New York. Contact: Len DeFrancisco.

United States Fish and Wildlife Service. Breeding surveys at Shawangunk Grasslands National Wildlife Refuge, 1998, 2001-2003. Unpublished data. Wallkill River National Wildlife Refuge, Sussex, New Jersey.

United States Fish and Wildlife Service. 2004. Waterfowl population status, 2004. U.S. Department of the Interior, Washington, D.C..

Usai, M.L. 1996. Region 9-Hudson-Delaware Report. Kingbird 46 (2): 183.

Chapter 4
What you can do to help protect IBAs

The protection and proper management of Important Bird Areas and the birds they support is a large, multifaceted endeavor requiring the efforts of many, from local citizens to conservation professionals to elected officials.

Audubon New York has made the conservation of IBAs a centerpiece of its mission and works toward that goal through its science, advocacy, and education initiatives. These include: research and stewardship projects for forests and grasslands; habitat restoration projects; promotion of IBAs in state and local open space protection plans; creation and proper management of state Bird Conservation Areas; and public education about priority birds and their habitats at Audubon environmental education centers and through other Audubon education programs. Often, Audubon takes the lead or acts alone on these initiatives. In many cases, though, we work in partnership with state and federal agencies, local Audubon chapters, land trusts, and grassroots advocates. We also play a supporting role for other conservation groups who have taken the lead in protecting specific habitats in their regions of the state. Our successes to date have been significant and many, but conserving IBAs is a big job, and we are always looking for more partners to reach our goal of protecting all of New York's IBAs.

Below is a list of some things you can do to help, in your home and community, through political advocacy at the state and federal levels, or through your profession. Whether you are a birder, work for a government agency or conservation organization, or do the shopping for your household, there are ways you can help protect IBAs and the birds that depend on them.

Your Home

Many important steps toward conserving birds and their habitats start at home with the actions of a single individual. The key is to stay informed and make good choices. To help, Audubon has established the Audubon At Home program, which provides homeowners with information and tools for taking the best conservation actions they can on their properties. Several suggestions are listed below, and more information can be found on the Audubon at Home web site (http://www.audubon.org/bird/at_home/index.html).

- **Make your yard a safe haven for migrating birds.** Eliminate the use of pesticides for aesthetic purposes and seek non-toxic alternatives. Many pesticides that control pests also kill birds. Also, protect birds in your yard from cats. House cats are non-native predators that kill millions of birds each year. Keep cats indoors and provide escape cover for birds such as shrubs and brush piles.
- **Avoid the use of invasive, non-native species.** Invasive species are one of the leading threats to birds and their habitats, including IBAs and National Wildlife Refuges. You can help stop their spread by not using exotic plants such as purple loosestrife, Japanese knotweed, and English ivy for landscaping in your yard, and by carefully cleaning your boat when taking it from one body of water to another. You can also support the efforts of the many groups that are monitoring invasive species populations,

providing for restoration of native species and habitat conditions in ecosystems that have been invaded, and promoting public education on invasive species and the means to address invasive species spread. You can maintain quality bird habitat and an attractive yard by choosing native plants that attract birds.

- **Be a bird-friendly consumer.** Careful choice in what products you buy can make a difference for birds. Purchase products for your home that are better alternatives for birds locally as well as globally. For example, buy shade-grown coffee to support coffee growing techniques that provide suitable winter habitat for many migratory birds; use forest products from companies employing sustainable forestry to help protect forest bird habitats; and buy organic produce to promote agricultural practices that are good for birds and good for the land.
- **Help monitor birds in your yard.** Participate in the Great Backyard Bird Count every February and be a part of a massive effort to provide an important mid-winter snapshot of bird distributions across the country while enjoying bird watching with your family. Other citizen science options that you can get involved in include the excellent offerings by the Cornell Laboratory of Ornithology, such as Project FeederWatch. For more information, contact Audubon New York or the Cornell Laboratory of Ornithology.
- **Be a steward of large tracts of land.** If you are the owner or manager of a large landholding, stewardship of your property may make a dramatic difference. If you are lucky enough to own property within an IBA, your land stewardship activities will have a direct impact on the habitat of priority birds. If you own and manage forests, work with a professional forester to manage the timber in a sustainable manner that supports significant bird species for the area. Contact Audubon New York for a copy of *Wildlife and Forestry in New York Northern Hardwoods: A Guide for Forest Owners and Managers*. Farmers with hayfields, pastures, and old (fallow) fields can contact Audubon New York for suggestions about accommodating birds in those habitats or information about enrolling in the Landowner Incentive Program or one of the conservation provisions of the Farm Bill. Both of these federal programs can assist private landowners in managing habitats on their property for the benefit of wildlife.

Your Community

- **Get involved with a local conservation group.** Join your local Audubon chapter or a similar group and help them adopt an IBA in your area to promote its protection and monitor its bird populations. Local groups can also participate in clean-ups or habitat restoration projects at IBAs, and they can lead public outreach and education efforts at sites where those activities are appropriate. As part of a local conservation group you can help your town or community be proactive on conservation by developing a local open space plan that identifies IBAs and other significant places that need protection. You can also initiate a Smart Growth movement to promote local development in an environmentally sensitive manner. Contact Audubon New York for information about starting a Smart Growth program in your community. Another important role of local groups is to serve as the eyes and ears for Audubon New York's IBA program, alerting us to activities that may be harmful to a site as well as opportunities for conservation actions. Finally, consider becoming a leader in your local group and encouraging others in your community to get involved.

- **Help monitor birds.** If you are a birder or want to be a birder, participating in bird monitoring activities in your community is an excellent way to help. You can never underestimate the power of good, up-to-date data. Data are important for assessing the significance of sites, monitoring population trends or distributions of birds, and guiding and evaluating conservation actions. Experienced and skillful birders can participate in surveys such as the national Breeding Bird Survey, Christmas Bird Counts, Winter Waterfowl Surveys, state Breeding Bird Atlases, and targeted IBA surveys developed by Audubon New York or local groups that have adopted IBAs. Even beginners can help monitor birds by participating in citizen-science bird monitoring programs such as the Cornell Laboratory of Ornithology's Birds in Forested Landscapes program. Finally, simply reporting your bird sightings in eBird, the on-line tool for collecting and organizing bird sightings developed by the Cornell Lab of Ornithology and Audubon, contributes to the efforts of conservationists and scientists.

Birders at May's Point, Montezuma National Wildlife Refuge

Political Advocacy

Some of the most important victories for birds and their habitats are won in the legislative and governmental arenas, and grassroots advocacy plays a crucial role. Whether conservation advocacy is new to you or not, a great way to stay informed of the important issues and take action is to join Audubon's Action Network. When you sign up for this free service, you receive the Audubon New York e-Nest and Audubon Advisory, as well as periodic action alerts when your help is needed the most. The e-Nest provides you with information about important conservation legislation before the state legislature and proposals being considered by state agencies, and the Audubon Advisory provides you with the best information on conservation legislation pending in Congress. Both also advise you on the best way to effectively communicate with your elected officials to make sure your voice is heard. To join or to learn more about these services, please visit: ny.audubon.org/action_network.htm or contact Audubon New York.

Listed below are some of the ongoing issues at the federal and state levels that are important for the protection of birds, wildlife, and their habitats.

Federal level

Ask Congress, including both your House of Representatives member and your two US Senators, to increase federal conservation funding, much of which is put to use at IBAs and other important habitats in New York. Some of the most important programs include the following:

- **Land and Water Conservation Fund** – provides money to federal, state, and local governments to purchase land, water, and wetlands for wildlife habitat and other uses.
- **State Wildlife Grants program** – enables the state wildlife agency to proactively work with public and private landowners on habitat conservation.
- **North American Wetlands Conservation Act** – provides matching funds to protect and restore wetland habitats that support birds and other wildlife.
- **Farm Bill** – contains many conservation provisions that assist farmers in conserving wildlife and wildlife habitat.
- **Landowner Incentive Program** – funnels funding to state wildlife agencies to work with private landowners to protect wildlife and their habitats.

In addition, there are numerous federal laws that are critical to maintaining the habitat quality at IBAs. Some examples to consider supporting include:

- **Endangered Species Act** – protects the most critically endangered species and their habitats.
- **Clean Water Act** – protects water quality across the country as well as wetland resources.
- **Clean Air Act** – addresses the issue of acid rain, which is harming aquatic and terrestrial ecosystems in New York State.

- **Migratory Bird Treaty Act** – protects birds from a wide variety of threats.
- **National Estuary Program** – protects and restores nationally significant estuaries like Long Island Sound. Also, support both the Long Island Restoration Act and the proposed Long Island Sound Stewardship Act, which benefit IBAs all around the shores of Long Island Sound by enhancing water quality and protecting ecologically significant habitats.

State level

Ask your Assembly Member, State Senator, and Governor to promote, enact and fund conservation programs that benefit birds, other wildlife, and their habitats. Promote the inclusion of IBAs in the New York State Open Space Conservation Plan, which makes them eligible for conservation dollars to either purchase the site in fee or conservation easement or enroll them in a conservation program for stewardship. As an individual, you can buy a Habitat Stamp (wherever hunting and fishing licenses are sold) to help fund non-game wildlife habitat management in New York. There are also opportunities to help wildlife programs through the voluntary state income tax check-off program and the state bluebird license plate program.

Governor Pataki (holding shirt at center) and David Miller (behind podium), Executive Director of Audubon New York, at the Montezuma BCA Dedication

As at the federal level, there are critical funding programs and laws at the state level that directly influence the conservation of IBAs. These include:

- **New York State Environmental Protection Fund** – provides for open space conservation and land acquisition.
- **New York Conservation Fund** – provides support to New York's wildlife conservation programs.
- **Smart Growth legislative initiatives** – if enacted, would protect open space and IBAs from sprawl development.
- **State environmental quality bond acts** – if passed, would increase funding in New York State for habitat protection.
- **Strong state wetlands protection, clean water, and clean air laws** – all help protect habitat quality in New York State.

Your Profession

Finally, many of you reading this book will be in a position to influence the conservation and management of IBAs directly through your civic duties or professions, whether you are members of your town council, natural resources professionals at an NGO or government agency, or elected officials. Here are some ways that key professions can contribute to the effort to protect New York's birds.

Town and county planners – Integrate IBAs into local zoning and development plans to protect these valuable habitats and maintain the quality of life for your citizens. Develop a local open space plan and launch an effort to promote Smart Growth.

Natural resource professionals – Help integrate habitat management for priority bird species into the management of properties that you oversee, especially those that are within IBAs.

Foresters and farmers – As outlined in the local community section, foresters and farmers can make land management choices that are good for forestry and agricultural objectives as well as for birds and their habitats. Become informed about new land use practices and programs. Contact Audubon New York as well as your county, state, and federal extension and conservation agencies to make a difference.

Land trusts – Target the IBAs in your area for acquisition or easements. Audubon New York staff would be happy to help develop management plans or reach out to landowners to discuss the significance of the habitats they own.

Elected officials – Take the lead on legislation and policies to protect IBAs and other significant habitats in your town, county, or in the state.

In the end, IBAs and the birds that depend on them will be protected only through the diverse actions of many individuals. Whatever your position or ability, please do your part to help protect New York's incredible natural heritage.

Appendices

Appendix A
Species at Risk Site Thresholds

The Audubon New York Technical Committee established site-specific thresholds for each species at risk based on the species at risk categories (see chapter 2), dispersion pattern (i.e. dispersed or aggregate), and taxonomic group (i.e. non-passerine or passerine). Site thresholds were used as guidelines to approve sites during the site review process.

Table 1. New York State Site Level Thresholds for Species at Risk

Species	*Risk Category*	*Threshold Breeding (number of pairs/number of individuals)*	*Threshold Migration (number of individuals)*	*Threshold Winter (number of individuals)*
Spruce Grouse	Severely at Risk	2/4	NA	6
Bald Eagle	Severely at Risk	2/4	15	15
Peregrine Falcon	Severely at Risk	2/4	6	6
Piping Plover	Severely at Risk	2/4	6	NA
American Oystercatcher	Severely at Risk	2/4	6	6
Roseate Tern	Severely at Risk	15/30	6	6
Least Tern	Severely at Risk	15/30	6	NA
Black Tern	Severely at Risk	5/10	30	NA
Black Skimmer	Severely at Risk	5/10	30	NA
Short-eared Owl	Severely at Risk	2/4	6	6
Sedge Wren	Severely at Risk	3/6	9	NA
Bicknell's Thrush	Severely at Risk	5/10	15	NA
Golden-winged Warbler	Severely at Risk	5/10	15	NA
Bay-breasted Warbler	Severely at Risk	4/8	12	NA
Cerulean Warbler	Severely at Risk	5/10	15	NA
Yellow-breasted Chat	Severely at Risk	3/6	9	9
Henslow's Sparrow	Severely at Risk	3/6	9	NA
Saltmarsh Sharp-tailed Sparrow	Severely at Risk	5/10	15	15
Seaside Sparrow	Severely at Risk	3/6	9	9

Species	*Risk Category*	*Threshold Breeding (number of pairs/number of individuals)*	*Threshold Migration (number of individuals)*	*Threshold Winter (number of individuals)*
Brant	Highly at Risk	NA	120	120
Common Loon	Highly at Risk	3/6	60	60
Pied-billed Grebe	Highly at Risk	5/10	15	15
American Bittern	Highly at Risk	3/6	9	9
Least Bittern	Highly at Risk	3/6	9	9
Osprey	Highly at Risk	5/10	15	NA
Cooper's Hawk	Highly at Risk	5/10	15	15
Northern Goshawk	Highly at Risk	3/6	9	9
Upland Sandpiper	Highly at Risk	5/10	15	NA
Red Knot	Highly at Risk	NA	120	120
Purple Sandpiper	Highly at Risk	NA	60	60
Short-billed Dowitcher	Highly at Risk	NA	120	NA
Common Tern	Highly at Risk	40/80	15	NA
Common Nighthawk	Highly at Risk	3/6	15	NA
Whip-poor-will	Highly at Risk	5/10	15	NA
Red-headed Woodpecker	Highly at Risk	3/6	9	9
Olive-sided Flycatcher	Highly at Risk	5/10	15	NA
Worm-eating Warbler	Highly at Risk	10/20	30	NA
Canada Warbler	Highly at Risk	10/20	30	NA
Grasshopper Sparrow	Highly at Risk	5/10	15	NA
Rusty Blackbird	Highly at Risk	5/10	120	120
American Black Duck	At Risk	10/20	240	240
Northern Harrier	At Risk	5/10	15	15
Sharp-shinned Hawk	At Risk	10/20	30	30
Red-shouldered Hawk	At Risk	5/10	15	15

Species	*Risk Category*	*Threshold Breeding (number of pairs/number of individuals)*	*Threshold Migration (number of individuals)*	*Threshold Winter (number of individuals)*
American Woodcock	At Risk	10/20	30	30
Willow Flycatcher	At Risk	20/40	60	NA
Horned Lark	At Risk	10/20	240	240
Wood Thrush	At Risk	20/40	60	NA
Blue-winged Warbler	At Risk	10/20	30	NA
Prairie Warbler	At Risk	15/30	30	NA
Vesper Sparrow	At Risk	15/30	30	NA

Appendix B
Nomination Form

Name of site: ______________________________ Submission Date: ______________

Audubon NEW YORK

IMPORTANT BIRD AREAS PROGRAM

Thank you for participating in Audubon New York's Important Bird Areas Program. Your help is greatly appreciated. Please read the accompanying Criteria and Guidelines before you fill out this form. Supply as much of the requested information as possible. Return this form to Jillian Liner, c/o Cornell Lab of Ornithology, 159 Sapsucker Woods Road, Ithaca, NY 14850. Please type or neatly print your entries. Return your nomination(s) by May 23, 2003. If you need further assistance, call the New York Important Bird Areas Coordinator at (607) 254-2437 or email at jliner@audubon.org.

Name: Last ____________ First ____________ Middle ____________ Title (if applicable) ____________

Organization (if applicable) ____________ Home Address: Street ____________ City ____________ State ______ Zip ______

Work Phone ____________ Home Phone ____________ email ____________ Fax ____________

SITE INFORMATION

1. Name of Site ____________ 2. County ____________ 3. Town ____________

4. General Location and Map

General boundaries of site: ____

____

____

____

Directions to site using nearest major town and highway: ______________________________

____

____

____

5a. Approximate Size in Acres ____________

6. Geographical Coordinates (center of site) Long ____ ____' ____" Lat ____ ____' ____"

7. Elevation Range (in feet) Lowest:______ Highest:______

5b. Accuracy of Acreage ____________

8. General Description of Site:

____

____

____

____

____

____

____

____

Nomination Form, continued

Name of site: ______________________________ Submission Date: ______________

ORNITHOLOGICAL INFORMATION

9. IBA Site Criteria - check all that apply. See Criteria for further details.

1. Sites where birds concentrate in significant numbers when breeding, in winter, or during migration.

- ☐ 1a. 2,000+ Waterfowl
- ☐ 1b. 100+ seabirds/10,000 Gulls
- ☐ 1c. 300+ inland/1,000+ Coastal Shorebirds
- ☐ 1d.100+ Wading Birds
- ☐ 1e. 8,000+ Raptors
- ☐ 1f. Exceptional Diversity
- ☐ 1g. Exceptional Concentration

☐ 2. Sites for New York State threatened species and other species of state conservation concern.

3. Sites supporting assemblages of species characteristic of a representative, rare, threatened, or unique habitat:

- ☐ 3a. rare, threatened or unusual habitat within the state or region.
- ☐ 3b. an exceptional representative of a natural or near-natural habitat within the state or region.

10. Ornithological Importance - this information is vital to supporting your site nomination. Please read the Guidelines carefully for instructions and fill in the table on page 3.

10a. Reference - please describe the references used to substantiate the data in the table on page 3 (one reference per line). Then fill in the Reference column of the table with the corresponding letter.

A ______________________________
B ______________________________
C ______________________________
D ______________________________
E ______________________________
F ______________________________
G ______________________________
H ______________________________

11. Additional Supporting Data - attach additional sheets if necessary

____
____
____
____
____

Nomination Form, continued

10. Ornithological Importance - this information is vital to supporting your site nomination. Please read the Guidelines carefully for instructions.

Criterion	Species/Group	Season (F, S, W, B)	Year	Average Number	Min Number	Max Number	Unit of Count	Length of Stay	# Years Present	Data Quality (1-3)	Reference

Nomination Form, continued

Name of site: ______________________ Submission Date: ____________

HABITAT INFORMATION

12. Habitats - classify as P (Primary: >50% of cover vegetation) or S (Secondary: <50% of cover vegetation) or provide approximate percent cover of each.

___Deciduous Woods
___Coniferous Woods
___Mixed Woods
___Shrub/Scrub
___Grassland
___Cultivated Field
___Marine
___Beach/Dune
___Tidal Wetland
___Non-tidal Wetland
___Lacustrine/Riverine
___Riparian
___Urban/Suburban
___Other ____________

13. Major Flora (especially rare and/or endemic):

Species	Importance of Site to Species

14. Non-avian Fauna (especially rare and/or endemic):

Species	Importance of Site to Species

15. Ownership - check all that apply. Please read Land Management/Land Ownership Information, page 6 of the nomination form, before completing.

___Private
___Corporate
___Municipal
___State
___Federal
___Non-governmental Conservation Agency
___Other ____________

16. Land Use(s) - classify as P (Primary) or S (Secondary). See Guidelines before completing.

___Agriculture/cultivation
___Pasture
___Water Supply
___Recreation/Tourism
___Mining
___Residential Development
___Commercial Development
___Urban/Industrial/Utility
___Forestry
___Fisheries/Aquaculture
___Wildlife Conservation/Natural Areas/Land Trusts
___Other ____________

Nomination Form, continued

Name of site: ______________________________ Submission Date: ______________

CONSERVATION

17. Threats - classify threats as: H (High), M (Medium), L (Low), or P (Potential).

___Non-native Fauna/Flora
___Over-extraction of Groundwater
___Pollution
___Cowbird Parasitism
___Succession
___Commercial Development
___Residential Development
___Infrastructure Development
___Recreation Development/Overuse
___Other ______________________
___Unknown ______________________

18. Summarize Threats to the Area:

19. Conservation Measures Taken, in Progress, or Proposed:

20. Economic, Cultural, and Social Values of the Site:

21. Additional Supporting Comments:

Nomination Form, continued

Name of site: ______________________________ Submission Date: ______________

LAND MANAGEMENT/LAND OWNERSHIP INFORMATION

An Important Note to Compilers (Read this before filling out the information below)

The IBA program is *not* a regulatory program, and a site's recognition as an Important Bird Area has *no* regulatory authority. The IBA program seeks first to identify habitats essential to birds, using a set of scientifically valid criteria. In all likelihood, an IBA designation will be indicative of good land stewardship and positive management practices that we want to encourage, with the full cooperation of the landowner or manager.

Audubon New York respects the rights of landowners and the right of privacy. Before making any final IBA selection, we will contact appropriate landowners and land managers to seek their consent and participation in the IBA process. Conservation objectives and alternatives for selected IBA sites will be set with the cooperation of the landowners and managers, as well as other interested parties.

We encourage compilers to approach landowners and managers when nominating IBAs *only* when there is a near-certainty of gaining the immediate understanding and cooperation of the landowner or manager, and *only* with the same respect that the compiler would expect for his/her own rights. A "letter of introduction" to landowners and managers is included with this packet.

If you have *any* doubts about how an IBA nomination will be received by a private landowner or manager, do not attempt to contact him/her. Instead, simply fill out the information below, if available, and indicate any potential problems in the "Comments" section.

OWNERSHIP INFORMATION

☐ Individual ☐ Corporate ☐ Municipal ☐ State ☐ Federal ☐ Other

Name of Owner 1

____

Name of individual contact, if applicable

____

Relationship of contact to owner

____

Address of owner

____

City

____

State ____________ Zip ______________

Telephone

____

Name of Owner 2

____

Name of individual contact, if applicable

____

Relationship of contact to owner

____

Address of owner

____

City

____

State ____________ Zip ______________

Telephone

____

Other comments/impressions (attach additional sheets if necessary):

____

____

____

____

____

____

____

Appendix C

A map showing the locations of IBAs within the Bird Conservation Regions of New York

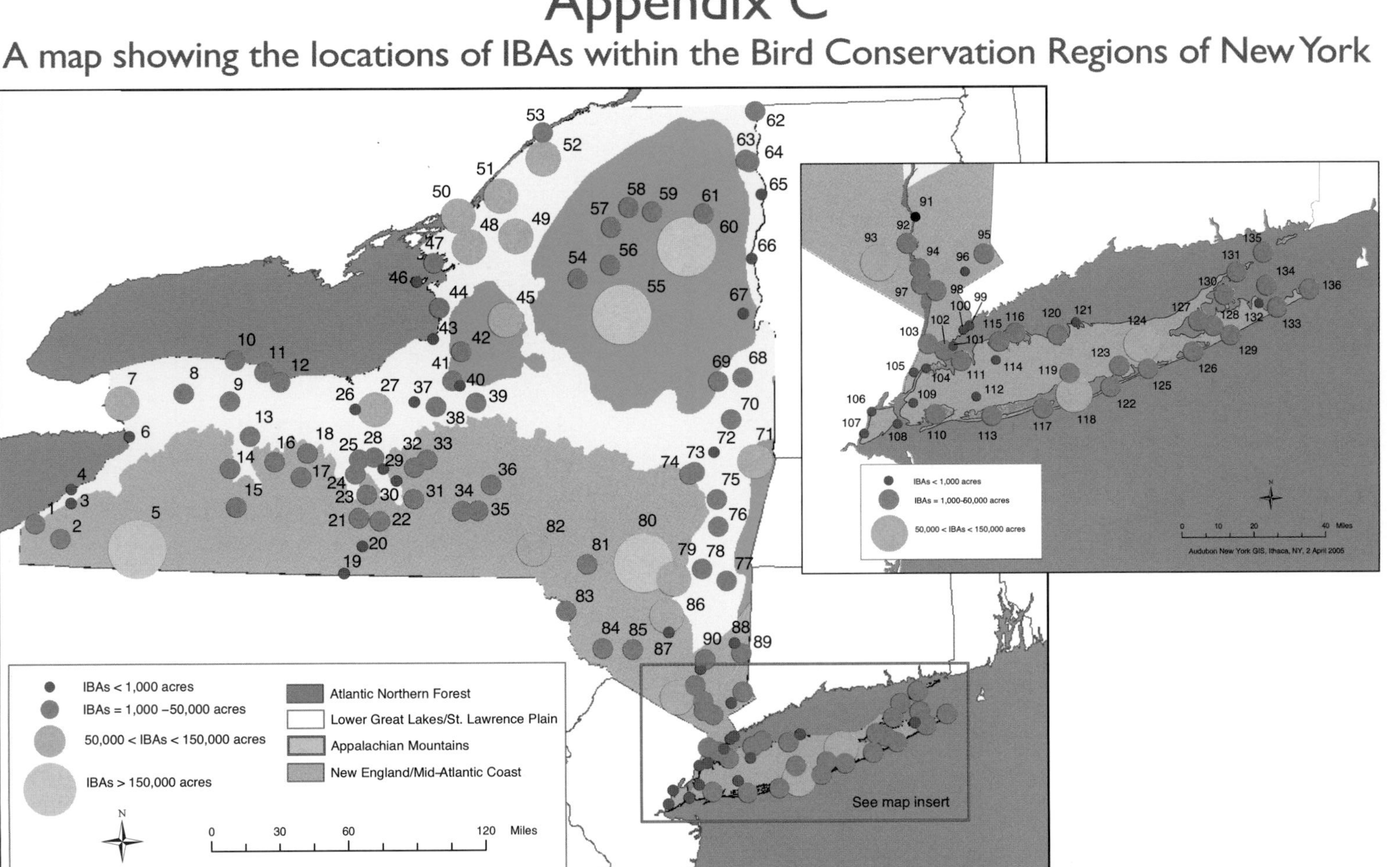

1 Ripley Hawk Watch (p. 96)
2 Chautauqua Lake (p. 50)
3 Wheeler's Gulf (p. 114)
4 Dunkirk Harbor/Point Gratiot (p. 58)
5 Allegany Forest Tract (p. 36)
6 Tifft Nature Preserve (p. 108)
7 Niagara River Corridor (p. 84)
8 Iroquois NWR/Oak Orchard and Tonawanda WMAs (p. 72)
9 Bergen Swamp (p. 40)
10 Hamlin Beach State Park (p. 64)
11 Braddock Bay (p. 42)
12 Rochester Area Urban Parks (p. 98)
13 Nation's Road Grasslands (p. 82)
14 Letchworth State Park (p. 76)
15 Keeney Swamp Forest (p. 74)
16 Hemlock and Canadice Lakes (p. 68)
17 High Tor WMA (p. 70)
18 Canandaigua Lake (p. 44)
19 Widger Hill (p. 116)
20 The Center at Horseheads Field (p. 106)
21 Queen Catharine Marsh (p. 94)
22 Connecticut Hill Area (p. 52)
23 Finger Lakes National Forest (p. 60)
24 Seneca Lake (p. 102)
25 Seneca Army Depot (p. 101)
26 Galen Marsh (p. 62)
27 Montezuma Wetlands Complex (p. 79)
28 Cayuga Lake (p. 48)
29 Aurora Grassland Complex (p. 39)
30 Salmon Creek (p. 100)
31 Caswell Road Grasslands (p. 46)
32 Greater Summerhill Area (p. 63)
33 Southern Skaneateles Lake Forest (p. 104)
34 Tioughnioga River/Whitney Point Reservoir (p. 110)
35 Long Pond State Forest (p. 78)
36 Pharsalia Woods (p. 92)
37 Whiskey Hollow (p. 115)
38 Onondaga Lake (p. 90)
39 Cowaselon Creek Watershed Area (p. 54)
40 Oneida Lake Islands (p. 88)
41 Toad Harbor Swamp (p. 112)
42 Happy Valley WMA (p. 66)
43 Derby Hill Bird Observatory (p. 56)
44 Eastern Lake Ontario Barrier Beaches (p. 130)
45 Tug Hill Area (p. 169)
46 Little Galloo Island (p. 146)
47 Point Peninsula (p. 162)
48 Perch River Complex (p. 158)
49 Fort Drum (p. 132)
50 Upper St. Lawrence/ Thousand Islands (p. 171)
51 Indian River/Black Lakes (p. 140)
52 Lisbon Grasslands (p. 144)
53 Lower St. Lawrence River (p. 148)
54 Stillwater Reservoir (p. 168)
55 Moose River Plains/ Blue Ridge Area (p. 152)
56 William C. Whitney Wilderness Area (p. 176)
57 Massawepie Mire (p. 150)
58 Spring Pond Bog Preserve (p. 166)
59 Adirondack Loon Complex (p. 124)
60 Adirondack High Peaks Forest Tract (p. 120)
61 Northern Adirondack Peaks (p. 157)
62 Chazy Landing/Kings Bay Area (p. 126)
63 Plattsburgh Airfield (p. 160)
64 Valcour Island (p. 174)
65 Four Brother Islands (p. 138)
66 Crown Point State Historic Site (p. 128)
67 Lake George Peregrine Site (p. 142)
68 Fort Edward Grasslands (p. 136)
69 Moreau Lake Forest (p. 156)
70 Saratoga National Historical Park (p. 164)
71 Rensselaer Forest Tract (p. 222)
72 Vischer Ferry Nature and Historic Preserve (p. 236)
73 Black Creek Marsh (p. 186)
74 John Boyd Thacher State Park (p. 212)
75 Schodack Island State Park (p. 226)
76 Stockport Flats (231)
77 Stissing Ridge (p. 230)
78 Tivoli Bays (p. 232)
79 Ashokan Reservoir Area (p. 182)
80 Catskill Peaks Area (p. 192)
81 Pepacton Reservoir (p. 221)
82 Cannonsville/Steam Mill Area (p. 190)
83 Upper Delaware River (p. 234)
84 Mongaup Valley WMA (p. 217)
85 Bashakill WMA (p. 184)
86 Northern Shawangunk Mountains (p. 218)
87 Shawangunk Grasslands (p. 228)
88 Little Whaley Lake (p. 214)
89 Great Swamp (p. 204)
90 Fahnestock and Hudson Highlands State Parks (p. 202)
91 Constitution Marsh Sanctuary (p. 196)
92 Doodletown and Iona Island (p. 198)
93 Harriman and Sterling Forests (p. 206)
94 Lower Hudson River (p. 215)
95 Ward Pound Ridge (p. 238)
96 Butler Sanctuary (p. 188)
97 Hook Mountain (p. 209)
98 Rockefeller State Park Preserve and Forest (p. 225)
99 Edith G. Read Wildlife Sanctuary (p. 200)
100 Marshlands Conservancy (p. 216)
101 Huckleberry Island (p. 210)
102 Pelham Bay Park (p. 306)
103 Van Cortlandt Park (p. 316)
104 North Brother/South Brother Islands (p. 296)
105 Central Park (p. 252)
106 Harbor Herons Complex (p. 268)
107 Clay Pit Ponds State Park Preserve (p. 254)
108 Hoffman and Swinburne Island Complex (p. 272)
109 Prospect Park (p. 308)
110 Jamaica Bay Complex (p. 276)
111 Little Neck Bay to Hempstead Harbor (p. 280)
112 Hempstead Lake State Park (p. 270)
113 West Hempstead Bay/Jones Beach West (p. 318)
114 Muttontown Preserve (p. 290)
115 Oyster Bay Area (p. 302)
116 Huntington and Northport Bays (p. 274)
117 Captree Island Vicinity (p. 246)
118 Great South Bay (p. 266)
119 Connetquot Estuary (p. 256)
120 Nissequogue River Watershed/ Smithtown Bay (p. 294)
121 Crane Neck to Misery Point (p. 258)
122 Fire Island (East of Lighthouse) (p. 259)
123 Carmen's River Estuary (p. 250)
124 Long Island Pine Barrens (p. 282)
125 Moriches Bay (p. 288)
126 Shinnecock Bay (p. 310)
127 Peconic Bays and Flanders Bay (p. 304)
128 Southampton Green Belt (p. 314)
129 Mecox Sagaponack Coastal Dunes (p. 284)
130 Northwest Harbor/ Shelter Island Complex (p. 297)
131 Orient Point/Plum Island (p. 300)
132 Accabonac Harbor (p. 244)
133 Napeague Harbor and Beach (p. 292)
134 Gardiner's Island (p. 262)
135 Great Gull Island (p. 264)
136 Montauk Point (p. 285)

Appendix D
List of the Audubon Chapters in New York and the IBAs intersected by their membership boundaries

Bedford Audubon Society
Butler Sanctuary
Great Swamp
Ward Pound Ridge Reservation

Bronx River/ Sound Shore Audubon Society
Huckleberry Island

Buffalo Audubon Society
Iroquois NWR/
  Oak Orchard and Tonawanda WMAs
Niagara River Corridor
Tifft Nature Preserve

Capital Region Audubon Society
Black Creek Marsh
John Boyd Thacher State Park
Rensselaer Forest Tract
Saratoga National Historical Park
Schodack Island State Park Area
Vischer Ferry Nature and Historic Preserve

Central Westchester Audubon Society
Edith G. Read Wildlife Sanctuary
Marshlands Conservancy

Chemung Valley Audubon Society
Cayuga Lake
Connecticut Hill Area
Finger Lakes National Forest
Keeney Swamp Forest
Queen Catharine Marsh
Salmon Creek
Seneca Lake
The Center at Horseheads Field
Widger Hill

Delaware-Otsego Audubon Society
Cannonsville/Steam Mill
Pepacton Reservoir
Pharsalia Woods

Eastern Long Island Audubon Society
Carman's River Estuary
Fire Island (East of Lighthouse)
Mecox Sagaponack Coastal Dunes
Peconic Bays /Flanders Bay
Southampton Green Belt
Long Island Pine Barrens
Moriches Bay
Shinnecock Bay

Four Harbors Audubon Society
Crane Neck to Misery Point
Long Island Pine Barrens
Nissequogue River Watershed/Smithtown Bay

Genesee Valley Audubon Society
Bergen Swamp
Braddock Bay
Hamlin Beach State Park
Letchworth State Park
Nation's Road Grasslands
Rochester Area Urban Parks

Great South Bay Audubon Society
Captree Island Vicinity
Connetquot Estuary
Fire Island (East of Lighthouse)
Great South Bay

High Peaks Audubon Society
Adirondack High Peaks
Adirondack Loon Complex
Chazy Landing/Kings Bay Area
Crown Point State Historic Site
Four Brother Islands
Massawepee Mire
Moose River Plains/Blue Ridge
Northern Adirondack Peaks
Plattsburgh Airfield
Spring Pond Bog Preserve
Stillwater Reservoir
Valcour Island
William C. Whitney Wilderness

Huntington Audubon Society
Huntington and Northport Bays
Muttontown Preserve
Oyster Bay Area

Jamestown Audubon Society
Allegany Forest Tract
Chautauqua Lake
Dunkirk Harbor/Point Gratiot
Ripley Hawk Watch
Wheeler's Gulf

New York City Audubon
Central Park
Clay Pit Ponds State Park Preserve
Harbor Herons Complex
Hoffman & Swinburne Island Complex
Jamaica Bay Complex
North Brother/South Brother Island
Pelham Bay Park
Prospect Park
Van Cortlandt Park

North Fork Audubon Society
Northwest Harbor/Shelter Island
Orient Point/Plum Island

North Shore Audubon Society
Little Neck Bay to Hempstead Harbor

Northern Catskills Audubon Society
Ashokan Reservoir Area
Catskill Peaks
Stockport Flats
Tivoli Bays

Onondaga Audubon Society
Cowaselon Creek Watershed Area
Derby Hill Bird Observatory
Eastern Lake Ontario Barrier Beaches
Happy Valley WMA
Oneida Lake Islands
Onondaga Lake
Toad Harbor Swamp
Whiskey Hollow

Orange County Audubon Society
Harriman and Sterling Forest
Lower Hudson River (incl. Croton Point Park)
Shawangunk Grasslands National Wildlife Refuge

Owasco Valley Audubon Society
Cayuga Lake
Greater Summerhill Area
Montezuma Wetlands Complex
Southern Skaneateles Lake Forest

Putnam Highlands Audubon Society
Constitution Marsh Sanctuary
Fahnestock and Hudson Highlands State Parks
Lower Hudson River (incl. Croton Point Park)

Rockland Audubon Society
Hook Mountain
Lower Hudson River (incl. Croton Point Park)

Saw Mill River Audubon Society
Lower Hudson River (incl. Croton Point Park)
Rockefeller State Park

South Shore Audubon Society
Hempstead Lake State Park
West Hempstead Bay/Jones Beach West

Southern Adirondack Audubon Society
Fort Edward Grasslands
Lake George Peregrine Site
Moreau Lake
Saratoga National Historical Park

St. Lawrence-Adirondack Audubon Society
Eastern Lake Ontario Barrier Beaches
Fort Drum
Indian River/Black Lakes
Lisbon Grasslands
Little Galloo Island
Lower St. Lawrence River
Perch River Complex
Point Peninsula
Upper St. Lawrence River

Sullivan County Audubon Society
Bashakill Wildlife Management Area
Mongaup Valley Wildlife Management Area
Upper Delaware River

IBAs not within a chapter boundary
Accabonac Harbor
Aurora Grassland Complex
Canandaigua Lake
Caswell Road Grasslands
Galen Marsh WMA
Gardiners Island
Great Gull Island
Hemlock and Canadice Lakes
High Tor WMA
Little Whaley Lake
Long Pond State Forest
Montauk Point
Napeague Harbor and Beach
Northern Shawangunk Mountains
Seneca Army Depot
Stissing Ridge
Tioughnioga River/Whitney Point
Tug Hill

Index